"It is I, Sea Gull"

Valentina Tereshkova, First Woman in Space

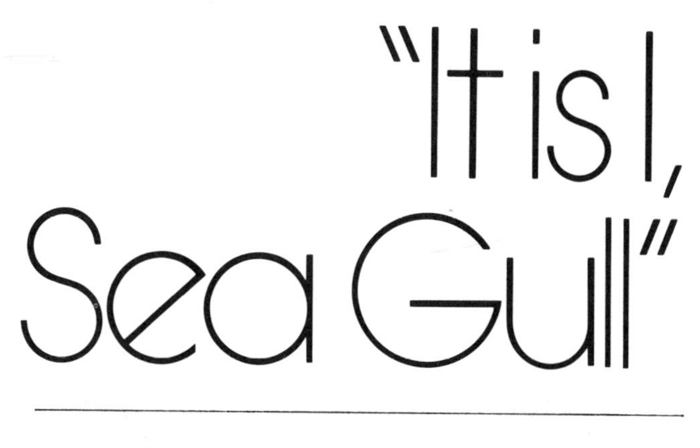

Valentina Tereshkova, First Woman in Space

BY MITCHELL R. SHARPE

ILLUSTRATED WITH PHOTOGRAPHS

THOMAS Y. CROWELL COMPANY NEW YORK

The publisher thanks the following for permission to use photographs: The British Interplanetary Society, London, and Cyril Maitland, *Daily Mirror*, page 161; Embassy of the U.S.S.R., Washington, D.C., page 187; George Hardy, Huntsville, Alabama, page 200; Marshall Space Flight Center, Huntsville, Alabama, page 193; Novosti Press Agency and Embassy of the U.S.S.R., Washington, D.C., pages 13, 33, 36, 38, 63, 72, 75, 83, 107, 125, 127, 130, 131, 134, 139, 150, 153, 163, 165, 166, 170, 173, 176, 180, 181, 192, 198; Mitchell R. Sharpe, page 191.

 Published simultaneously in Canada by Fitzhenry & Whiteside Limited, Toronto.

Designed by Angela Foote

Manufactured in the United States of America

Library of Congress Cataloging in Publication Data
Sharpe, Mitchell R. "It is I, Sea Gull."
SUMMARY: A biography of the first woman astronaut and history of the Russian manned space program. Bibliography: p. 205. 1. Nikolaeva, Valentina Vladimirovna Tereshkova, 1937- —Juv. lit. 2. Astronautics—Russia—History—Juv. lit. [1. Nikolaeva, Valentina Vladimirovna Tereshkova, 1937- 2. Astronauts—Russia. 3. Astronautics—Russia—History] I. Title. TL789.85.N48S5
629.45'0092'4 [B] [92] 74-14698 ISBN 0-690-00646-2

1 2 3 4 5 6 7 8 9 10

This book is dedicated to the young women of the world today who, in a few years, will follow Sea Gull into space.

Acknowledgments

After having written *Yuri Gagarin, First Man in Space,* it was natural that I should turn to a biography of Valentina Tereshkova, the first woman in space. This book was begun in the days before the current détente with the U.S.S.R., and the material available in the U.S.A. was limited severely. It was not until 1971 that I began to acquire material from the Soviet Union not available generally in the United States. In that year, I was invited by the Academy of Sciences of the Soviet Union to read a paper on the history of rocketry in the United States at the Twenty-third International Congress for the History of Science, held at the State University in Moscow.

During the several weeks I spent in Moscow, I met many Soviet scientists and historians who expressed a desire to assist me in gathering material for my proposed biography of their famous astronaut. In particular, Dr. V. N. Sokolsky, chairman of the National Association of Historians of Science and Technology of the Academy of Sciences of the Soviet Union, was most cooperative. Since my return to the United States, he has continued to assist me greatly. In addition, A. Skripkin, curator of the Tsiolkovsky State Museum of the History of Cosmonautics in Kaluga, whom I also met in Moscow, has been most generous, as has been A. V. Kostin, grandson of the famous Russian pioneer in astronautics, Konstantin E. Tsiolkovsky. Mr. Kostin was an excellent host at the Tsiolkovsky Home Museum, also in Kaluga, during my visit there in 1971. And, of course, I am equally grateful to Vladimir S. Vasiliyev, my friend at the Institute of Studies of the U.S.A., of the Academy of Sciences of the Soviet Union. He has been my research assistant in Moscow during the preparation of the book.

I also have to thank various members of the staff of the Embassy of the Soviet Union for supplying endless photographs, many of which have been used in this book.

—Mitchell R. Sharpe

Contents

A Sea Gull Among the Swallows

It was bitterly cold in the tiny hamlet of Maslennikovo on March 6, 1937. Snow lay on the ground among the bare birch and ash trees of the forest surrounding it. A single early swallow greeted the dawn.

In a few months, however, the peasants of the Priziv collective farm would hear thousands of swallows singing that spring had returned.

In those days, Maslennikovo was a collection of some forty rude huts, resembling the log cabins of the settlers who had populated the New World two centuries earlier. Clinging to the top of a small hill some 190 miles northeast of

Moscow, and surrounded by several ponds, Maslennikovo had neither electricity, paved streets, nor running water. Indeed, the women and children of the hamlet had to carry water from a well at the bottom of the hill in buckets on a yoke over their shoulders.

In one of these huts on that day, Vladimir Aksenovich Tereshkov, a mechanic and tractor driver on the collective farm, waited nervously. His wife, Yelena Fedorovna, was expecting their second child, and the baby was clearly coming early.

Vladimir Aksenovich was a tall and husky man with curly blond hair and gray-blue eyes. He worked hard, and when time permitted, he relaxed by playing a squeaky accordion. But this was no time for the accordion. He was worried about Yelena. Despite his solicitude, he was pushed outside the hut by his mother, who was helping ensure that her grandchild would arrive in this world safe and sound.

The new baby inherited her father's blond, curly hair and gray-blue eyes as well as his height. Yet she bore the traces of her mother's dark beauty, too.

Yelena Federovna had decided before her second daughter was born that she would be named Rayechka Vladimirovna, but her strong-willed husband had another name in mind. Without telling her, he went to the district council and registered his new daughter's name as Valentina.

When a child is christened in the Soviet Union, it takes its father's first name as a middle name. For boys, the ending *vich* is added to it, while the ending *ovna* is added for girls. Thus, Valentina's middle name was Vladimirovna. The Russians use both the first name and the middle name when addressing or referring to someone for whom they have great respect.

When Valentina's father returned to Maslennikovo, he told his wife that their daughter Valentina Vladimirovna was now a citizen of the Soviet Union. Then there occurred one of the few quarrels that Valentina's parents ever had.

Valya, as her family called her, soon became her father's favorite. Her mother seemed to prefer Ludmila (Lyuda), her older sister. But Valentina's memories of her father were to be few. She would remember one spring day, though. She was four years old, and the sun was beating down on the fields of the farm. Her father scooped her up suddenly and placed her on the seat of the tractor. Before she knew what was happening, the machine lurched forward with a roar and clatter. She was terrified by the acrid smoke and the noise.

The young man told his daughter to put her hands on the steering wheel so she could help him drive. With her hands on the wheel helping him steer, Valya became self-confident. Vladimir Aksenovich kept telling her what a fine job she was doing. Soon Valya began to smell the richness of the soil turned up by the plows and to notice the blue skies and white clouds above the rolling hills. Most of all, though, she was fascinated by the birds—freed from earth—sweeping high above and well beyond the Volga River.

One afternoon not long afterward, Valya returned to the hut to find Mama and Grandmama in tears. Soon Lyuda, who was three years older than Valya, joined them in crying. Their father was going away with the army. On June 22, 1941, the Germans had attacked the Soviet Union. It was hard to explain to Valya what war was or who Germans were. There was no war in Maslennikovo. There were no soldiers and cannon, no bombs and killing. She could not understand why her father had to leave her. She also could not understand why Mama, Grandmama, and Lyuda were crying so much. In fact, Valya did not even know what *leaving* meant.

Vladimir Aksenovich never came home. A letter did

arrive, and it said that he had died for the homeland and been buried as a hero. Valya did not know what all of the words meant, but when Mama, Grandmama, and Lyuda all burst into tears, she too started crying.

For Valya, her father became only a yellowish picture made by a lonely soldier in a penny arcade and sent home to his family. He looked down at her from the mantel, a solemn young man in a drab uniform. With his wire-framed glasses, he looked more like a schoolteacher than the tractor driver who had given her a ride in the fields of Maslennikovo.

Sometimes, on seeing the picture as she grew older, Valya would say to her grandmother, "What kind of a man was my papa?"

"Oh, an ordinary person, just like everyone else," she would answer. "Except he was a little stronger and more handsome than other men. You, Granddaughter, are the image of him."

Then Grandmama would add that had Vladimir Aksenovich lived, he probably would have become the chairman of the Priziv farm, because everyone looked to him for guidance and valued his opinion.

Only a few months after Vladimir Aksenovich died, Valya and Lyuda were hurried out of the hut by their grandmother, just as she had hurried their father out on two previous occasions. When Lyuda and Valya were allowed back into the hut, Mama was lying in the bed with a very tiny red baby. He was brother Vladimir (Volodya). This time there was no doubt about the baby's name! Mama named him Vladimir Vladimirovich, for his father. She personally entered his name in the records of the district council.

As the war went on, life became worse for Yelena Fedorovna and her three young children. The men of the collective farm helped her as much as they could, but it soon became obvious that Lyuda and Valya would have to pitch in also.

One of the first things the girls learned to do was to look after Volodya. However, Volodya proved to be more than they could handle. He weighed too much for them. Unable to carry him about, Lyuda and Valya let him crawl on the floor. From time to time, Volodya would let out screams as he picked up splinters from the rough planks of the floor. When Mama came home, tired from working in the fields, she was greeted by Volodya crying and Valya and Lyuda arguing. Then Mama herself would begin to cry.

To add to Mama's troubles, one day Valya and Lyuda decided to make some clothes for Katyusha, their favorite doll. They chose what they thought was an old rag, but it turned out to be Mama's only winter jacket.

In despair and frustration, Yelena Fedorovna screamed, "I am going to leave all of you and go away into the world!"

Valya and Lyuda really began to cry. The world, they knew, was somewhere on the other side of the far-off city of Yaroslavl.

The expensively clad Katyusha had a strange fate. The girls secretly included her in a package of clothing and food the people of Priziv sent to the soldiers. When they later heard that "Katyusha" was causing terror among the Germans, they were proud but mystified as to how their doll could do such things. It was not until after the war, when Valya was studying history, that she learned Katyusha was the name of a rocket used by the Russians against the Germans.

It was not long before Lyuda and Valya realized that they had to settle down. The war which they had not earlier understood suddenly came home to Maslennikovo. Into the tiny village arrived refugees whose homes had been destroyed and who were fleeing from the Germans. These people came from Leningrad, that fabulous city which Lyuda and Valya knew only as one of those impossibly faraway places like Moscow.

Among these refugees, Valya and Lyuda got to know two girls, Raya and her sister, also named Valya. These two girls told them many stories about the famous city in which they lived. Now their home in Leningrad had been destroyed by the bombers which flew over daily and by the long-range cannons which shelled it day and night.

Learning to work was not easy. One of the first things the girls tried to do was to help milk the cows on the collective farm. It looked easy enough. Lyuda sat down on a stool behind their cow Tamarka. She reached up confidently enough and grasped at two of the teats and pulled down as hard as she could. Tamarka immediately kicked with both back feet, sending Lyuda in one direction and the milk bucket in another, and terrifying Valya.

Hearing the noise in the barn, Grandmama rushed in. She explained that nothing in life was as simple as it seemed to be—milking was one of those things. Then she showed the girls how a cow was milked.

The girls also learned how to carry water from the well at the bottom of the hill. By balancing the yoke across their shoulders they could carry the two buckets on the ends fairly easily. However, they found that they had to walk carefully in order to keep the buckets from swaying too much. In the summer, Valya and Lyuda covered the buckets with cabbage leaves to keep the water cool as they trudged up the hill.

Another task was taking the noon meal to their mother, who worked in the fields. The way through the thick fields of rye was long and dusty, and the sun was blistering. When the workers caught sight of five-year-old Valya and eight-year-old Lyuda, not too much taller, trudging through the high grass carrying the lunch bucket wrapped in a kerchief, they would call out, "Look, Yelena, here comes your crew."

Summer meant there was time for fun, too. When she was six years old, Valya and her friend Tonya Moshova learned to swim in one of the ponds near the village. Valya delighted in climbing to the top of a large birch tree and plummeting into the pond, a feat the boys would not attempt.

The children of Maslennikovo also played "war," and inevitably the Red Army won, with the captured German prisoners locked in an outhouse. When Valya and her friends got tired of this game, they made up others. One was particularly prophetic. Valya and her friend Tonya had quite a reputation among the village children as tomboys. Tonya was more of a sister even than Lyuda in many ways. Indeed, Yelena Fedorovna had nursed Tonya when she was a baby, and Tonya's mother had nursed Valya. Valya also felt especially close to Tonya because she, too, had lost her father in the Great Patriotic War, as World War II is called in the Soviet Union. It was not surprising that they were the inventors of a game called "paratroopers." It consisted of climbing into tall birch trees, crawling out along the supple branches, and descending slowly to the ground as the flexible limbs bent under their weight.

Like real paratroopers, the two tomboys often went home with torn clothes, skinned knees, or bruised noses. Sometimes, despite these wounds, Mother would inflict still more for Valya's daring in playing the forbidden game again after having been warned against it.

To add to the family's troubles, Yelena Fedorovna became seriously ill and was taken to the hospital in Yaroslavl. Now the three children had to leave their home and move in with their grandmother. Life was harder still because Grandmama had little time to spend with Lyuda, Valya, and Volodya. She had to work on the collective farm, too.

And the war continued to go badly. Daily the women of the village received letters from their husbands, sons, and other soldier relatives. They hated to see the postman arrive with his collection of letters, folded up into triangles and bearing no stamps, as was customary for letters from the front. From time to time, some of the village men would return to Maslennikovo. They were always bandaged, and some would be without an arm or a leg. Others would be blind, led by comrades.

Everyone chipped in to help Grandmama and her three small grandchildren. Though food was extremely scarce, neighbors brought in what they could spare from their own tables. Uncle Arkady, chairman of Priziv, would bring them a loaf of rye bread when he could manage. Others donated potatoes and some milk. They lived on these simple foods for many months.

During the winter, when the snow threatened to bury the small hut in which they lived, the three children and their grandmother slept on a bed on top of the warm stove. To conserve the precious kerosene, the lamp could not be lighted every night. To help pass those long, dark nights, Grandmama would tell traditional Russian fairy tales, such as the one about the firebird and the hunchbacked horse. In the cozy warmth and under the spell of Grandmama's crooning voice, Valya, half-asleep, would imagine that she was riding the hunchbacked horse across the skies. Sometimes she would see her mother as the beautiful Vasilisa and Uncle Arkady as Ilya Murometz, the victorious nightingale bandit of the story.

On days when it stopped snowing for a while and the sun peeped out, Valya would join Tonya in sliding down the hill on a sled made from an old basket top. Half the fun was in trying to miss the hemlocks and oaks on the way down.

In mid-June 1943, Valya and Lyuda knew that the German bombers were attacking Yaroslavl, only forty miles to the north, on the Volga River. They were trying to knock out the rubber-tire factories, textile mills, and diesel-engine plants in the large city. Mama had told them that more than 300,000 people lived in Yaroslavl. They were all afraid that their own grandmother, who lived there, would be killed.

However, there was a happy event to distract them. Mama came home from the hospital!

With the Tereshkov family settled back in its own hut, things grew better. Lyuda began school, and she came home each day with something new to tell Valya. As Lyuda sat beneath the kerosene lamp reading her lessons, Valya was awed by the fact that those curious little squiggles could spell words like *mother* and *bread*. When there was enough kerosene for the lamp after the homework, Lyuda and Valya would slowly thumb through the pages of an old copy of *Niva* magazine, which had been around the hut for so long that even their mother had forgotten where it came from.

When Valya was eight years old, the Great Patriotic War ended. The end of the long war marked a new phase in her life. She was to leave the tiny village of Maslennikovo and move to Yaroslavl. For over a year, Yelena Fedorovna's mother had been trying to get her and the children to move into Yaroslavl. Her house was on the outskirts of the city, and there was plenty of room for all of them in it.

Borrowing one of the few farm trucks left by the

summer of 1945, Yelena Fedorovna, with the help of neighbors, loaded the furniture from the hut on it and sent it ahead. She and the three children followed on foot, with Tamarka the cow and Murka the cat. For a long way they traveled on paths among the broad farm fields. From time to time they stopped to let the cow graze and drink from a pond or stream, and to eat something themselves. Valya's job was to take care of Murka, who for most of the journey slept curled up in a little bucket that Valya carried.

Everything went well until the Tereshkovs reached the highway that turned left to Yaroslavl and right to Uglich.

Turning left on the highway, the Tereshkov parade came upon a great steel bridge. It was the first one that Valya had ever seen and it was frightening. Tamarka refused to cross it and lay down in the middle of the highway. Valya put down her bucket to help Mama and the other children prod Tamarka onto her feet and push her across the bridge. While they were busy with this task, Valya heard a noise behind her. The bucket tipped over, and Murka ran for the woods bordering the highway. Valya and Lyuda ran after her, but it was useless. Murka made it into the woods and disappeared.

The two girls immediately began to cry despite Yelena Fedorovna's assurance that Murka was not lost forever. Indeed, she was right. A week later, Murka turned up back at Grandmama's hut in Maslennikovo, and she lived there for years afterward. Grandmama said that Murka had run away because she did not want to live in Yaroslavl.

About three miles from Yaroslavl the railroad crosses the highway. Just as the Tereshkovs got to it, the guard came out of his house and slammed down the gate across the highway. With a roar, the train whipped past them in a cloud of steam. The huge locomotive was pulling a string of green passenger cars, the windows of which blended into one long silver ribbon as the train sped by.

"Mama, what's that?" Valya screamed.

Yelena Fedorovna was still worried about the cow, who had never seen a train before either.

"What's that? Oh, Valya, it's a train," she said, tugging on the rope around Tamarka's neck.

"But there was a man in the front, leaning out of the window," Valya went on.

"He's the engineer, the one who drives the train," she said as the guard lifted the gate.

Once the train was out of sight, the family pushed, pulled, and prodded poor Tamarka across the tracks. Valya's fear had quickly left her and a firm conviction had taken its place. On the spot, she made up her mind that she was going to be a locomotive engineer when she grew up.

Grandmother's house was not actually in Yaroslavl proper. It was in a suburb called M.V. Frunze, and it was on a street named Strike. By Maslennikovo standards, the house was a mansion. Grandmother lived in one half of it, but the other half was vacant. It was filled with dust and junk, but in a matter of days Yelena Fedorovna had it spotless.

Valya was amazed to find her new hometown so big and yet so old. It had been founded in about 1010 by the Grand Prince Yaroslav of Kiev, who was also known as Yaroslav the Wise. World War II was not the first time Yaroslavl had been destroyed. As early as 1238 it was burned by the Tatars, and in 1332 it was burned again by the Grand Duke Ivan Kalita. Each time, however, the people immediately set about rebuilding the city—just as they were doing in 1945 when Valya became one of its newest citizens.

Summer passed quickly for Valya as she was meeting new friends and getting acquainted. She slowly overcame her homesickness for the peace and quiet of Maslennikovo. Instead of the brilliant blue skies she was used to, the skies of Yaroslavl were black with the smoke from mills and factories. The trees outside her grandmother's house did not seem to whisper, and she missed the song of the nightingale at evening.

But there was one good thing. Not far from the house ran the railroad. Valya could sit by the hour watching the trains go by, their engineers smiling and waving to her. Valya's preoccupation with trains and locomotives became an obsession with her and a bore to her friends. Whatever the topic of conversation, Valya steered it around to trains. One day she announced that she was going to become an engineer.

"*You* drive locomotives?" said one of her friends sarcastically.

"I will, I will . . ." muttered Valya.

Valya found an ally in her determination to become a locomotive engineer: Anya Pelevina, who lived next door and was a switchman on the railroad.

"Why shouldn't Valya become an engineer?" Anya Pelevina said to Yelena Fedorovna.

But Yelena Fedorovna was equally determined that Valya should *not* become a locomotive engineer.

Valya listened with excitement as Anya Pelevina read to her from a story called "Signal." It was about something that a crossing guard named Semen Ivanov did to prevent a railroad wreck. As Anya Pelevina read, it seemed to Valya that she could actually see the red and green semaphores, hear the shrieking whistle . . .

". . . He took off his cap and pulled out a cotton handkerchief," she read. "He pulled out a knife from his boot, crossing himself and asking for the Lord's blessing. Just above his left elbow he slashed his arm, and the blood gushed out in a hot stream. He soaked the handkerchief in his own blood and tied it to a stick, which he waved. . . ."

In the fall of 1945, eight-year-old Valya entered Primary

Valya, third from the right, in the second row from the back, with her classmates in Yaroslavl.

School No. 10 in Yaroslavl. Mama got her up early that day. She woke her singing:

> *"Children, get ready for school!*
> *The rooster has crowed long ago.*
> *While you yawn, get dressed.*
> *The sun is looking in the window."*

Then Mama did Valya's hair up in two braids with white ribbons. The school satchel was checked to make sure that books, paper, pencils, were in it. Then they joined the procession of other first-graders and mothers heading toward the two-story stone school.

Valya's teacher was Mariya Mikhailovna Kokoreva, who had not long been out of teachers' college. In fact, Mariya Mikhailovna lived on Strike Street not far from the Tereshkovs. However, Valya soon found out that having the teacher live so close by was not always a good thing. If Mama or Grandmother happened to be home after school, Mariya Mikhailovna would often stop for a chat and quite often might mention some of the pranks that tomboy Valya had performed in school that day.

The school itself was a new adventure. Every room was crowded with pictures, many of them depicting famous events in the Russian revolution such as "The Storming of the Winter Palace." There was also a statue of Lenin on an armored car during the revolution. In addition, there was a large collection of dried plants and flowers as well as an equally large collection of stuffed animals and birds. Both of these collections had been made by the older children of the school over the years.

Valya's favorite subjects were arithmetic and Russian, but she also liked art. While her classmates drew heroic pictures of battles or workers in the factories, Valya always turned to the same subject: Maslennikovo. She was still a little homesick for the village that now seemed so far away. The picture was always of the hut she had grown up in beside the tree-lined ditch. Next to it, in the yard, was Mama. In one picture she had on a yoke and was returning from the well. In another she had a bundle of firewood on her back. In still another she was washing clothes.

With the coming of spring, Valya found herself promoted to the second grade. During the summer holiday, she spent part of the time in the Young Pioneer Camp, which was like an American Girl Scout camp, and the rest of the time back in Maslennikovo with Grandmama.

After a year or so in Yaroslavl, Valya had become completely a city girl. Her entire way of life had changed. Indeed, the whole family's way of life had changed in ways of which they were scarcely aware. Yelena Fedorovna no longer baked bread, not even the little buns they all loved so much. She sent Valya or Volodya to the bakery for them. All their food was bought now in the stores, and for the first time Valya could depend on having candy more than once a year, as had been the case in Maslennikovo. Mama did not mind if she took a few kopeks from the change to buy gumdrops or ice cream. Money had become more plentiful when Lyuda had finished school and started to work in the Red Perekop textile mill, where Mama already worked.

Homework was much easier now also, since Valya studied under electric lights rather than a kerosene lamp. It seemed natural now to board a bus or a streetcar to go across town or on a holiday outing. Only a year or so earlier, she had never heard of such transportation. In the same way, she regarded the radio and movies as something she seemed always to have known.

However, there was one thing about her that did not change. She remained a tomboy.

Just as she had in Maslennikovo, Valya found it hard not to accept a challenge from a boy. She was always ready, on a bet, to run (not walk!) through the cemetery at night.

Another challenge was the tall concrete bridge over the Kotorosl River near the factory where Yelena Fedorovna worked. This river was the one in which Valya and her friends in the Young Pioneers swam in the summer—and she was, of course, one of the first in her crowd to swim across it. However, only the oldest and bravest boys would jump from the bridge into the river.

"Why are the boys any better than us?" Valya asked several of her girl friends one day as she squinted into the sun while looking up at the bridge. "It's only a jump," she added with a grin.

Some boys overheard her remark and immediately challenged her. The group of children rushed onto the bridge and looked over the railing and down into the churning river beneath it. Despite her attempts to look calm, Valya was scared stiff. It was such a long way down! Not even the top of the tallest birch tree over the pond in Maslennikovo was as high as this bridge. But Valya knew she could not back down.

She climbed up on the railing and closed her eyes tightly. The next thing she knew, she was rushing through the air—for hours it seemed. Then, suddenly, the coldness shot through her like an electric current. When she bobbed to the surface, she looked up at the bridge. It was crowded with people, grownups as well as her friends and the boys who had dared her to jump. She swam quickly to the bank, with teeth chattering, and was soon back among the crowd on the bridge. In a few minutes she did something that really thrilled them. Valya clambered up on the rail and jumped again!

When her head popped above the surface, everyone on the bridge was cheering and applauding.

Soon other girls joined her in jumping. On her last jump of the day, one of the boys told her to spread her arms out and keep her head forward while she was in the air. She did.

On her way to the water she heard him calling that she looked just like a sea gull.

When Yelena Fedorovna heard about the feat, she could only shake her head.

"It would have been better if you had been born a boy," she said.

Following elementary school, Valya entered Girls' High School No. 32. She was assigned to fifth class "C," which was her homeroom. The homeroom teacher also taught French, and it was all she thought about. Her desk was stacked with novels and books of short stories and poems by Balzac, Stendahl, Mérimée, Rolland, Barbusse, and Aragon. These books, like the ones she constantly carried about in her arms, were filled with impromptu bookmarks: a flower, a hairpin, a scrap of paper, or a bus ticket.

For geography, Valya had a teacher who was especially talented in making the subject interesting. One day, while resting with his hands on the globe, he looked quietly at the class and said, "The time will come when man will take in the whole earth at a glance just as we now are looking at this globe here in our classroom."

Valya's history teacher was particularly interested in the history of Yaroslavl. She did more than merely lecture on the subject—she took Valya and her classmates on field trips. They visited the home of the poet Nikolai Nekrasov, and then they memorized his poem about the Volga River. Together they trooped over the whole city, studying the ancient buildings and fortresses that went back to 1237, seven hundred years before Valya had been born.

Mathematics was another matter. This teacher was strict and permitted no nonsense. She went by the book. Her pupils only once made the mistake of coming to class without having finished their homework. The unfortunate soul who did come unprepared was made to stand by the blackboard throughout the period. Luckily for Valya, mathematics was not hard for her. In fact, she helped her classmates over some of the rough spots in algebra and geometry, although it meant studying together until late at night. However, the cooperation paid off. They all got through the course with passing grades.

If mathematics was a chore, literature was a pure pleasure. It was taught by a pretty, dark-haired young woman who walked quietly and proudly like the heroine in one of the Russian novels or plays she taught. Valya and her classmates often thought of her as the tragic Anna Karenina of Leo Tolstoi's novel or as Mikhail Lermontov's beautiful Tamara, who died so tragically after being kissed by the Demon.

Valya's reputation as a tomboy preceded her to high school. She was made class monitor to ensure that other students did not get unruly or into trouble. Being the monitor taught her to control her own urge to cut up.

As Valya's high school days drew to a close, she and her closest friends held long, serious talks about what they wanted to do when they graduated. Her mind was still made up. She wanted to be a locomotive engineer. In particular, she wanted to enroll in the school of railroad technology in Leningrad.

Graduation day, in the spring of 1953, was curiously like her first day in elementary school. Mama made sixteen-year-old Valya a brown woolen dress and white pinafore, and she plaited her hair into two long braids, each with a white ribbon.

Aleksandra Fedorovna Ilyasova, the principal, handed out the diplomas and looked the girls over carefully. "The straight road is the best, and those who do not turn from it will go far," she said solemnly.

The joy of graduation did not last long once Valya got home.

Yelena Fedorovna put her foot down in no uncertain terms. Absolutely not. It was ridiculous. Valentina Vladimirovna Tereshkova was not going to Leningrad to study to be a locomotive engineer.

"What kind of an engineer would you make?" she said scornfully. "You couldn't even pull the throttle. You're just a little bird . . ."

Valya thought of all the reasons why she should go. She even made a muscle in her arm to show that she could pull the throttle. She begged Lyuda to help her out.

"No, Valya, you should go to work in the mill and study at night school," was her sister's only reply.

Sea Gull Tries Her Wings

A sixteen-year-old just out of high school in Yaroslavl faced the universal problem of finding a job. College of any kind was out of the question for Valya; she couldn't afford it. She began trudging from factory to factory, filling out endless employment applications. Usually the frustrating comment she received was, "Come back when you have a little more experience, young lady."

At the end of August 1953, Valya gave up job-hunting and enrolled in the Young Workers' School No. 10 to study some technical subjects that she had not taken before. It was a night school, and she found herself in a strange and depressing atmosphere, quite different from that she had known in high school. Instead of girls her own age for classmates, Valya had a collection of older strangers. Some were women her mother's age, with daughters as old as, or older than, Valya. Sitting at the same desk with her was Boris Klimenko, a locomotive fireman, old enough to be her father.

Recesses were different, too.

The men went outside to smoke. The women talked of problems at home or at work. Some of them, too weary for gossip after working all day, simply closed their eyes and rested against the wall or bowed their heads on their desks, catching a few winks before the next lecture.

Valya wondered if this was to be her fate. Would she, too, wind up working in a factory all day and continuing her education at night like these people? Would she be able to stand the grind? Somehow she felt she was destined to do something better. Talking with Boris kept alive her dream of eventually becoming a locomotive engineer. Perhaps, someday, some way, she could get to Leningrad and the school there.

Near the end of June 1954, Valya finally found a job. It was one that no teen-ager could be overjoyed in getting. She became a "stripper" in the Yaroslavl Order of Lenin Tire Factory, in Department No. 5, where the tires were made. From the beginning, things got off to a bad start. She was assigned to the first shift and told to report to work on the next morning. That night she was so excited that she simply could not go to sleep until very late. She was still asleep when Yelena came in from her late-night shift at the Red Perekop mill.

She was surprised and angry to find that Valya was still

soundly asleep. She began shaking her and yelling that she had only a half hour to get to work.

Thirty minutes! The tire factory was almost an hour's ride away. Valya was going to be late for work on her first day! She broke into tears. Without stopping for breakfast, she ran for the streetcar and swung aboard it.

With eyes red from crying, Valya took a seat. Gradually, she began to blush. She was sure, by the way they were looking at her, that all the passengers were sneering. Actually, none of them noticed the pretty blond girl huddled in the seat next to the window.

The trip was agonizingly slow. The streetcar stopped at every street crossing, and it took hours to load and unload passengers. It moved like a snail down the tracks toward the edge of town. The driver seemed to move in slow motion as he closed doors, let off the brakes, and pushed the control forward. It was almost as if he were taking pleasure in going as slowly as he could, knowing Valya was becoming later and later.

The foreman was waiting for her, hands on hips and a scowl on his face. Valya burst into tears. The scowl on the foreman's face slowly softened, and he gruffly told her she had better buy an alarm clock.

A "stripper" in a tire factory works with the fabric base on which the rubber tread is applied. As the piles of rubber-coated fabric emerge from a machine, the stripper cuts them to length and readies them for vulcanizing. It was hard work and demanded a certain skill. If it had not been for Vera Nikolayevna Posokhova, Valya would have required much more time to master her job. Vera had worked in the tire plant most of her life, and knew each job perfectly.

"Don't be afraid to handle the cord, Valya," she said. "Whack it off, and fit it into place; don't take so much time."

Under Vera's instruction and encouragement, Valya gradually gained confidence and skill.

Her first paycheck came at the end of the month. It was a small one because she had not worked a full month, but it was the first money she had ever earned. With it, she decided to buy a small gift for her mother: a kerchief and a box of candy.

"Mama, I got my first paycheck," Valya said as she ran into the house with the gifts.

She was completely surprised at Yelena's reaction. Her mother hugged her and then burst into tears! Valya could not understand why. Neither could her uncle and aunt, who were visiting.

"What's the matter, Yelena, you ought to be happy to have another wage earner in the house," they said.

Yelena did not explain, but she went to a mirror and tried on the kerchief and smiled. Then she made a samovar of tea, and the family shared the box of candy.

As Valya became more skilled, her pay grew, and she took home more money. With three people working, the burden on Yelena Fedorovna lightened. The family could afford a few luxuries. One of these was a radio. It soon became almost a necessity. The family gathered around it at night and listened to the news, and to operas and other musical programs. Valya, who had always liked music, became addicted to opera. Her favorites were Tchaikovsky, Glinka, and Beethoven.

The tire plant was hot and noisy. Even though the windows were open, and Valya's machine was near one, the heat of the place was stifling. Through the window, and above the clatter of the machines, from time to time there came the whistle of a locomotive as a train chugged past. At such times Valya would pause for a moment, and the ever-present smell of hot rubber would be replaced by the smell of burning coal and hissing steam. She would no longer be a "stripper," but an engineer with one hand on the throttle and the other on the whistle.

The dream would soon be shattered by the smell of hot rubber and clattering machines, and a yell from the boss!

Work in the tire plant became even more of a burden on Valya when she had to rotate shifts. When her turn came for the night shift, it meant that she had to miss school because she could not always find someone to trade shifts with her. In addition, her studies were becoming harder, and she spent most of the time on the overcrowded streetcar trying to read. The pace became too much.

In April 1955 Valya transferred to the Red Perekop Factory No. 2, where her mother and sister worked, as a winder in the ribbon department. It was only a short distance from her home, so there was no long streetcar ride to and from work. She walked to the factory with Mama and Lyuda.

The textile plant was different in many ways from the tire factory. There were more machines and fewer people. Also, the machines were quieter, and there was no ever-present, sickening smell of hot rubber. But the work was more exacting. It required a quick hand and eye that the stripper did not need. The winder had to keep an eye on 126 spindles and take them off when they were full. It meant being in constant motion. Eyes, hands, and feet had to work together perfectly. However, Valya soon learned her new trade.

Indeed, within a year, she was tending two machines at once! At Red Perekop she worked on the piecework system. The more she produced, the more she was paid. At first her pay was very small, but as she gained skill, it increased. One of the first things Valya bought was a bicycle for Volodya on his birthday. He kept it for years, later riding to work on it after he had finished high school.

After completing the course in night school, Valya entered the secondary technical school that was associated with the textile factory to further her education. She would work hard for five years before graduating in 1960. Again, it was a night school, so she worked all day and went to classes at night. It left her with little free time for dates, but by then she was used to such social deprivations.

The new courses were tough. They included metallurgy, machine operation, electrical engineering, methods of testing cotton fabrics, and principles of textile manufacture, to name but a few. One of the teachers was Yevgeny Pavlovich Shcheglov.

"I envy you girls," he once said. "You have everything going for you; you will get ahead and become the big bosses in this industry."

The girls smiled at the thought that they might someday become bosses of their factory. However, they knew why their teacher was stressing the value of a technical education in the rapidly expanding economy of the Soviet Union as it rebuilt after World War II. Certainly women would occupy managerial positions, because so many men had been killed in the war. Strangely enough, Valya was not excited by the idea of becoming a captain of industry. She still had her dream of being an engineer of a locomotive.

Just as things were going well, misfortune struck the Tereshkovs in 1957.

Yelena Fedorovna became seriously ill and had to quit work. The burdens of the family fell on Lyuda and Valya. They had to shop, cook, keep house, get Volodya off to school, nurse their mother, and work at Red Perekop as well. It soon became apparent that Valya could not meet all her responsibilities and keep up in night school as well. She thought seriously of giving it up so she could nurse her mother. But she stayed in school. The increasing difficulty of the courses required her to sit up long hours into the night reading and studying while she kept an eye on Mama.

In January 1957 Valya joined the Komsomol (Young Communists' League) at Red Perekop and proudly pinned on her small red badge with Lenin's profile. She at once became active in its recreational and cultural activities. Being especially fond of music, she played the mandolin in a folk-music group. On Sundays and holidays Valya joined the other members in visits to historical sites, motorboating on the Volga River, and attending the ballet or the movies. But her participation with the group was restricted because of all the other demands on her time.

By the end of spring, Yelena Fedorovna was feeling better, and Valya took a vacation trip to Leningrad to visit relatives and friends. It was the first time she had ever been to a really large city. For years she had studied about Leningrad, the fabulous city of the old czars built on a swamp by Peter the Great in 1703. Like all tourists, foreign or domestic, she was enchanted by it.

Leningrad is also a city important in the history of rocketry. It was here in the early 1930's that Valentin Petrovich Glushko, the designer of engines for the rocket booster that orbited *Sputnik 1* and the Vostok manned spacecraft, began his career. His work was done in the famous Gas Dynamics Laboratory. Later, significant research in rocket propulsion was performed by the Leningrad Group for the Study of Reactive Motion, founded by Nikolai Alekseyevitch Rynin and Jakov Isidorovich Perelman. As early as 1933, this group in Leningrad designed, built, and static-fired a liquid-propellant rocket engine that produced ninety-one pounds of thrust.

Valya strolled behind the walls of the gloomy Peter and Paul Fortress, situated on the small island, Zayachie, where the Little Neva separates from the mainstream. Among the other things she noted was a dark, damp tower in which the guide said Nikolai Ivanovich Kibalchich had been imprisoned. When one of Valya's group asked who he was, the guide explained that he was a revolutionist of the Narodnaya Volya (People's Freedom) party, which had made the bomb that almost assassinated Czar Alexander II on March 1, 1881.

While Kibalchich had been locked in this tower, he had drawn up a sketch for a manned rocket ship. He had left it with his jailers before his execution. However, it was not discovered until 1917.

Dutifully, as befitted a member of the Komsomol, Valya trooped aboard the cruiser *Aurora,* permanently anchored in the Neva River not far from the log cabin in which Peter the Great lived his first eight years in the city. This ship played a pivotal role in the Great October Socialist Revolution in 1917. The ship's crew, on November 7, joined with the revolutionists of the Red Guard and trained the ship's guns upon the Winter Palace that housed the offices of the provisional government. (At the time of the revolution, Russia was still using the old-style calendar. Thus, the "October" revolution occurred in November as we date events today.)

The thing about St. Isaac's Cathedral that impressed Valya most was its tower and spiral staircase. From the top she looked out over the city. To the north she saw the Neva with its bridges and the golden spire of the old Admiralty Building. There also, tiny as it appeared, she could make out the statue of Peter the Great called "The Bronze Horseman." To the east she saw the Mars Field, with the Marble Palace at one end and the Mikhailovksi Palace (now the Russian Museum) at the other.

The view from St. Isaac so fascinated her that Valya stayed until well after dark, watching the lights of the city come on. A chill blew in from the Baltic Sea and brought Valya back to reality. Reluctantly, she trudged back down the staircase and returned to her relatives' apartment.

It took a whole day to go through the Hermitage, which is one of the greatest art museums in the world. Many of its paintings had once belonged to Catherine the Great. Most of its priceless collection of modern masters, however, had been involuntarily donated by Sergei Shchukin and Ivan Morozov, two prerevolutionary millionaire businessmen of Moscow. Shchukin was particularly fond of the French impressionist Henri Matisse, and thanks to him the Hermitage had such paintings as Matisse's "Nasturtiums with 'the Dance.' "

The Picassos, Renoirs, Van Goghs, Gauguins, Cézannes, Matisses, and Monets bothered Valya. They had none of the "socialist realism" she had been taught to admire in art. They did not relate to human problems and situations. In none of them did she see a tractor or a crane. Cézanne's "Mont Saint-Victoire" really did not look to her like a mountain.

Valya was particularly interested in the many Greek and Scythian art objects, some of which dated back to the fifth century B.C.

Even though she spent a whole day at the Hermitage, Valya saw only a fraction of its 2 million objects.

She felt much more comfortable and at home in the Russian Museum, which displays only Russian paintings and art objects. She admired Karl Brynllov's "The Last Days of Pompeii" and Orest Kiprenski's "Portrait of Colonel Evgraf Davydov" because she had studied about them in school. However, it was Ilya Repin's famous painting "The Volga Boatmen" that captured her attention. Its realism caused her to feel immense sympathy for the peasants tugging on the line attached to the heavy boat in the river.

Time in such an enchanting place as Leningrad was a precious commodity. Knowing that she might never again visit that great city, Valya was determined to squeeze the most out of every minute. Since she was there at the end of spring, during the period of the famous "white nights," when the sun sets very late because the city is so far north, Valya was still strolling through the streets and along the famous canals in the evening. Despite the gloomy rain that drizzled down through the "white nights," they were every bit as romantic as the Russian poets Pushkin and Blok had made them out to be. For that matter, she romantically recalled as she walked alongside the Griboyedov Canal, Dostoyevsky had titled his best work on love *White Nights.*

Like all vacations, Valya's was over before she realized it. She had to return to drab, smoky Yaroslavl and to her work at Red Perekop.

Her friends in the factory and the Komsomol gathered around her and questioned her. Many of them had never been to Leningrad and were familiar with it only through books and movies.

"Did you go to Petrodvorets and see the fountains?"

"Oh, yes! I saw them!"

"Valya, how do the girls dress in Leningrad?"

"Just like we do."

"Did you see many sailors on the Nevsky Prospekt?"

"More than we see in Yaroslavl."

Then there came a question that brought a deep blush to the cheeks of Valentina Vladimirovna Tereshkova, leading member of the Red Perekop Komsomol and future secretary of the organization.

"Valya, tell us about Lenin's home at Razliv."

With flaming cheeks, Valya had to say that it was the one famous place in Leningrad that she had not visited!

Valya soon settled back into the routine of Red Perekop. As the fortieth anniversary of the Great October Socialist Revolution approached, the citizens of Yaroslavl were caught up in the excitement of the holiday. The workers of Red Perekop were planning special events.

However, all these events were overshadowed by one planned several years previously by a group of Russian scientists who had experimented with crude rockets in the 1920's and 1930's in Moscow and Leningrad.

On October 4, 1957, the Soviet Union launched *Sputnik 1,* the world's first artificial satellite. It was an aluminum sphere 22 inches in diameter and weighed 184 pounds.

Valya and her fellow workers at Red Perekop were extremely proud to learn that their country had been the first to send a man-made object into orbit about the earth. That evening, at night school, her teacher said, "Just imagine, for the first time a creation of man's hands is in outer space!" After a moment of silence he went on, "This is the first step on the way to the stars."

While Valya, and indeed the world, was still in awe of *Sputnik 1,* the Soviet Union launched the world's first living creature into orbit, the dog Laika. On November 3, 1957, a rocket with a capsule containing Laika was launched from Tyuratam, the Cape Kennedy of the Soviet Union, a place Valya would come to know sooner than she could imagine.

Valya was immensely proud that her country had made these two important contributions to science and technology, but she felt no motivation to become a part of the space effort. She was far too busy finishing up her technical training and doing her practical work or internship. In 1960 she took her final examinations and presented her thesis. After receiving her certificate as a textile-engineering technician, she was transferred from the ribbon department of her plant to the mechanical-repair department. Thus her knowledge of textile machinery was put to better use.

However, Valya's tenure in this new job was brief. She was elected to the full-time position of secretary for the Red Perekop Komsomol. In this job she had to plan and manage the activities and programs of twenty organizations within the factory. One of her most memorable projects was a Komsomol outing to Moscow and a visit to the Economic Achievements Park. She was especially interested in the exhibits of textile machinery. However, she also looked with curiosity and pride at several objects in the Cosmos Hall. There was a replica of *Sputnik 1;* a model of a spaceship designed by K. E. Tsiolkovsky, the father of astronautics; and models of early rocket motors built and tested in the Soviet Union during the 1930's.

"Valya, look here!" one of the girls called to her. "It's a space cabin like the one for Laika," she went on.

The group quickly clustered around the model of Laika's life-support system and read the placard next to it which told how air and food were supplied to the dog during her orbits of the earth. As Valya joined the others and read how the complex system operated, she saw what a feat of engineering the system was. Her technical training in textile machinery gave her a good appreciation of it. She was beginning to take a more personal interest in her country's fledgling space program now that she could see tangible evidence of it and appreciate the skill in engineering that was going into it.

Her appreciation was also heightened by her own involvement in parachuting, which had begun in the fall of 1958 when her friend Galina (Galya) Shashkova, who had been in technical school with her, began urging her to join the Yaroslavl Air Sports Club. "You could learn to parachute," Galya said. "Of course, some people are afraid to take a chance . . ."

These words were not lost on the tomboy who had been the first girl to jump from the bridge over the Kotorosl.

The airport at Yaroslavl, where the club was, at that time was a smooth green field on the edge of town. On one side ran the railroad, and on the other was the highway down which barefooted Valya with her mother, sister, brother, and the cow had come thirteen years earlier.

Valya joined Galya on her next trip to the club.

Members of its parachuting group, dressed in their blue flight suits, were buckling themselves into the green parachutes. They took little notice of Valya when Galya left her to put on her suit and parachute.

When Galya reappeared, she and the others climbed aboard a Yak-12 airplane. The plane roared off the runway, banked steeply to the right, and began spiraling up over the airport. Valya squinted against the sun and followed it as it climbed to 5,000 feet. Soon bundles began falling from the Yak-12's open door, and parachutes bloomed against the blue sky. As the jumpers drew near the ground, Valya could see how they were guiding themselves by pulling first on the front and then on the rear risers. One of them landed only twenty feet from her, and quickly gathered in the chute and collapsed it. Valya was surprised to see that it was Galya, and she was equally surprised to see how professionally Galya handled herself as a parachutist.

"Well, you see how easy it is, don't you?" said Galya with a slightly superior tone.

Valya was convinced. She joined the club on the spot. However, it was a long time before she took to the air. Her instructor was a fellow worker at Red Perekop. He was a methodical man, and he began at the beginning. Before he

The new parachutist in her flight suit.

went into the theory and description of the parachute, he recounted its history. He showed the novice class a sketch of a parachute drawn by Leonardo da Vinci in 1514, and he then traced the development to the exploits of Russian parachutists such as Vasily Romanyuk, who was the first man to make 3,000 jumps.

After the lectures came six months of ground training and learning to pack parachutes. It was all very interesting, to be sure, but the class was fidgeting. Valya and the others wanted to start jumping.

"When are we going to get to jump?" asked Tanya Torchilova, an art student and a new friend of Valya's.

"After a light rain on Thursday," the instructor replied with a grin as he gave his standard answer to impatient students.

Curiously enough, May 21, 1959, *was* a Thursday, and it began with a spring shower. Valya climbed into the Yak-12 with Tanya, who seemed not so impatient, and the other members of the novice class who were to make their first jump. The pilot pulled the plane up sharply and banked around to begin the spiral up to 5,300 feet. Valya checked her main chute and her reserve continuously. In her mind she went over every step in the procedure that she had been taught.

The droning of the plane's engine had almost a hypnotic effect on Valya. In one way it was good. It calmed her considerably. But suddenly she had an urge that she could not explain. It seemed as if someone had said, "Valya, jump." Compulsively she moved to the door, automatically checked her chute, attached the static line, took the proper position, and leaped from the plane. The rest of the class was amazed!

As she plummeted toward the ground, for just an instant there pulsed through her the feeling she had experienced many years earlier when she had jumped from the bridge over the Kotorosl River. Now, however, instead of a river rising up to meet her, it was the earth. Quite clearly below she saw the Kotorosl and the bridge, Red Perekop, and, probably only in her imagination, her home near the factory. Suddenly there was a crack like a rifle shot, and Valya felt as if she had hit the ground. Her chute had opened.

The rest of the voyage to earth was almost like a dream. She floated down, automatically pulling at the risers to keep her feet pointed back toward the airport. The hangars and planes grew larger and larger. Suddenly, the ground itself was upon her. She landed lightly, running and collapsing the chute exactly as she had been shown time and again. Her first jump was over. This was the first of 125 jumps she would make in her career. She was extremely proud of herself, and grinning broadly as she turned to meet the instructor, who was running toward her.

"What do you mean by jumping before the command was given?" he screamed at her. "The first thing you have to learn in this game is to follow orders," he went on. "I am almost tempted to throw you out of the club."

She had never seen him like this. And she was terrified. Tears began to form.

"Well, I'll overlook it this time," he said. "But you must learn from this mistake. Follow orders."

Only after the jump did Valya tell the family about her new sport.

She was late returning home. "Where have you been, Valya?" Volodya said. "Mama and Lyuda have been worried about you."

Valya hurriedly explained what she had been doing and asked her brother to intercede for her with Mama. He did, explaining that practically all of her friends' daughters were members of the club.

"Jump, my child, but not too often," Mama said sadly.

The Yaroslavl Air Sports Club in 1960. Valya is on the right in the second row.

The next morning Yelena Fedorovna still had her mind on Valya's parachuting. "She's come a long way from parachuting from birch trees in Maslennikovo," Yelena Fedorovna mused.

It was another month before Valya made her second jump. The third followed in only a week or so. It was a very important one. On that jump Valya had to use her reserve chute, to teach her what to do in case the main chute failed to open. These specialized jumps continued through June and July of 1959.

By the middle of July, Valya and the other novices were permitted to take part in a competition with another club. On her ninth jump Valya leaped from an altitude of 2,640 feet, performed perfectly, and landed with precision in a bull's-eye on the ground. She got a grade of "excellent" and had her picture taken with two other young women, both named Valya. It appeared in the local newspaper, and Yelena Fedorovna cut it out and placed it in her scrapbook.

Three months after her first jump, Valya had done so well that she received the rating of parachutist third-class. But it took much hard work and many jumps before she would be awarded her first-class rating two years later.

As Valya participated more in the flying club and in parachuting, she became increasingly aware of the rapid advances being made in space exploration by her country. On January 2, 1959, *Luna 1,* man's first probe to the moon, was launched. It was during this time, too, that Vasily Romanyuk made his record-breaking three thousandth parachute jump. In an interview with a reporter from *Pravda,* Romanyuk said, "And who knows, perhaps the time will come when the world's first cosmonauts will also return to our planet under the canopy of a parachute." This statement by one of her idols made a very deep impression on Valya.

The pace of activity in the flying club stepped up, and Valya had little time for thoughts of space flight. She had reached the phase of training that required her to perform a free fall with a delayed chute-opening. This type of jump is extremely dangerous, and it requires a great deal of skill and self-control. It simply is unnatural for someone with a parachute to leap from an airplane and fall like a rock toward earth without pulling the ripcord. It also is very difficult to fall gracefully and not somersault or "corkscrew." Valya's first efforts went badly. It seemed she simply couldn't work out the timing. The instructor encouraged her, but she still could not perform to her own high standards and expectations. Valya seriously considered quitting the club, but she could not face the alternative: admitting defeat. She stuck it out.

Still, it was rough going. During a district competition, she went into her free-fall jump and immediately began spinning. However, she managed to straighten out—but she landed clumsily. It was too much for her instructor.

"Valya, you're about as graceful as a bear!" he said in disgust.

Valya picked herself up and ran off to the edge of the field where no one could see her. She sat heartbroken and crying among the daisies, pulling the petals off one of them. Then she heard someone running up behind her. It was Tanya.

With Tanya's alternate scolding and encouragement,

A scene from the Russian documentary film Road to the Stars. *Valya and a fellow parachutist climb aboard their small plane.*

Valya prepared for the next jump. This time she was determined to do better, and if not, at least she would not break into tears like a child. She made a jump in perfect form and landed only sixty-five feet from the target! This time the instructor was the first to meet her, and he gave her a hug. Tanya winked at her, and Valya clasped her hands above her head like a boxer who had just won by a knockout.

But Valya could not spend all her time parachuting. The duties of secretary for the Red Perekop Komsomol organization made many demands on her time.

On April 12, 1961, Valya had just started to hold a Komsomol meeting in one of the shops. She had opened the proceedings, but before she could get down to business, a girl raced in.

"There's a man in space!" she yelled. "He's a Russian and his name is Yuri Gagarin!"

The room burst into applause and sudden chatter.

Some of the chatter became ridiculous, at least in Valya's opinion.

"I wonder what color eyes he has," one girl said.

Another sang an answer.

"Don't fall in love with dark eyes;
dark eyes are dangerous;
but do fall in love with blue eyes;
blue eyes are beautiful. . . ."

Valya tried to go on with the meeting, but it was useless. So she gave up. Everyone was sitting around the radio listening to the details of Gagarin's historic orbit of the earth. Now his ship, *Vostok 1,* was over South America, and Gagarin reported that everything was going all right. Over Africa, a few minutes later, he reported that weightlessness was no problem for man in space. Finally, at 10:55 A.M., only 108 minutes after launch, he landed beyond the Volga at a place called Smelovka, southeast of the town of Engels.

Finally Valya joined the rest of the Komsomol in a spontaneous parade through the streets of Yaroslavl. She helped compose a telegram from the organization to the world's first cosmonaut.

Later Valya and her family heard over the radio that Chairman Nikita Sergeyevich Khrushchev had said to the young cosmonaut, who had just circled the earth, "Not only your parents but also our entire Soviet motherland is proud of your great triumph, Yuri Alekseyevich. You have achieved a triumph which will live for ages."

When the excitement over *Vostok 1* had died down a little, Valya began to think of the flight in feminine terms. Before this time, she had not seriously thought about a woman cosmonaut. But she was greatly impressed by the fact that the world's first cosmonaut was a Russian and that he came from the ordinary people like herself. He was certainly no superman. He was a short, blond fellow with a square face and a friendly grin—and he had a wife named Valentina!

Shortly after Gagarin's mission, his book, *The Road to the Stars,* appeared. The first few printings sold out as quickly as they hit the newsstands and bookstores. But Valya managed to get a copy. As she read it, she could not completely put aside the idea that a woman could do all these things in space as well as a man. However, she did not think that it would be someone like her.

Still, she could not get over the fact that Gagarin devoted several pages in his book to the importance of parachuting as a part of his training to be a cosmonaut.

She remarked on this one day at the flying school. In fact, she went so far as to suggest that all of the members should strive to jump as if they were cosmonauts. Cosmonauts!

The instructor sarcastically referred to her as "Gagarin in a dress," and she did not mention female cosmonauts around him again.

Nevertheless, as secretary of the Red Perekop Komsomol, she booked the various groups to see a film called *The Maiden Voyage to the Stars,* which was showing at the Motherland Theater in Yaroslavl. It was a movie that interested her because, for the first time, it revealed the equipment on which the cosmonauts were trained. Valya went to the movie several times.

On one occasion Volodya and Lyuda went with her.

"They really put those guys through a wringer, don't they?" observed Volodya in awe.

"I don't think that a woman could stand up under all that stress," commented Lyuda demurely.

Valya listened to these comments. Her technical training gave her a good idea of what the various machines were meant to do. She also knew that women are not necessarily the weaker sex. Indeed, she had read somewhere that doctors suggested that women would be better as cosmonauts than men because they consume less oxygen! There was only one thing that bothered her. It was the centrifuge. She simply did not know whether a woman could stand up under the stress of the high accelerations imposed by the machine. At an acceleration of 6 g's, she would "weigh" 630 pounds instead of her normal 105 pounds!

Later, as Yuri Gagarin made his world tour after the historic *Vostok 1* mission, collecting honors in every country, Queen Elizabeth II of England asked him if *women* would someday join men in space.

"*One* surely will," said Gagarin. "In our country there is equality between men and women, so . . ."

These words were not lost on Valya, but she still had no *personal* feeling of commitment to space.

At the Komsomol meetings the inevitable question

came up: Can a woman go into space as a cosmonaut? The debates were long and passionate, usually with the women on the yea side and the men on the nay side.

"Well, it probably won't happen soon," said one of the young women.

"In any case, outer space is nothing for girls in a textile factory to think about," added another.

"Besides," said still another, "they probably have already selected a woman cosmonaut—and she is a young graduate of a university in one of the sciences or engineering, or perhaps she is a doctor of medicine."

The same discussion began to disrupt the proceedings of the Yaroslavl Air Sports Club.

The instructor sought to break it up. "All right, you female cosmonauts, let's get down to parachuting," he said.

Valya immediately wondered why he had said that. Was it merely to motivate his students, or was there another reason? She knew that he was well respected in Moscow for his ability to turn out excellent parachutists.

Space would have to wait. Valya had more important matters to think about. The celebration of the USSR Air Force Day was approaching, and the Yaroslavl Air Sports Club was scheduled to put on a big show. It would not be nearly as big as the one at Tushino Airport in Moscow, but it would be really big by Yaroslavl standards. Valya and her fellow members brought it off in style. There was stunt flying and precision parachute jumping, and the crowds surrounding the airport were fully satisfied.

Valya's part in the show was a first for her. And what a first! She had to jump into the Volga River!

As she treaded water waiting for the motorboat to pick her up, she remembered that it was only a year earlier that she had been in the boat that picked up the club members who were making their first water jump. While she did not realize it, Valya had mastered a step in parachuting that would be very important to her future.

Several weeks later, on Sunday, August 6, 1961, Valya was at home. She was listening to an opera on the radio before starting to work on the schedule for the Red Perekop Komsomol for the coming week. The opera was interrupted by an announcement:

Vostok 2 had been launched into orbit, and the cosmonaut was Gherman Titov, who had been the backup for Yuri Gagarin.

Titov orbited the earth for a full day in *Vostok 2.*

It was evident to Valya that manned space flight was here to stay. At the same time, she wondered if there would also be "womanned" space flight.

She sat down and wrote a letter to the Supreme Soviet in Moscow asking that she be considered for cosmonaut training.

The result? Nothing. No word after a week, after two and then three weeks.

It was obvious to Valya that she had acted foolishly and impetuously. She actually blushed when she thought about the letter she had written. The scientists in Moscow must certainly be laughing at the naive letter from a textile worker in Yaroslavl stating that she would like to be a cosmonaut!

What Valya did not know was that something very important to her future was going on in the inner councils of her country's government.

The twenty-second Congress of the Communist Party of the USSR was meeting in the new Hall of the Congress within the walls of the Kremlin. One of its major concerns was expanding the role of the Soviet Union in space exploration—a role in which it was already supreme.

The decision was made to include women in the

nation's space program, and there was much discussion among the officials as to where such Russian women could be found. Obviously there were few women who had flown jets in the USSR. But a number of Russian women did have pilot's licenses or were qualified parachutists. Since the manned program called for the returning cosmonaut to leave his spaceship and descend by parachute, this latter group looked inviting to the space-program planners in Moscow.

Valya's letter had arrived at an opportune moment.

However, she had more or less forgotten about becoming a cosmonaut, and she threw herself into her parachuting and Komsomol work. She was working toward her "jubilee" jump, the one hundredth. However, it was not a matter of numbers only. Each jump was a prescribed one and had to be done to the satisfaction of the instructors before it counted. While she was working toward her "jubilee," Valya also became a public instructor and began teaching pupils of her own, fellow workers from Red Perekop.

One day in the early fall, Valya and the others learned that a committee of very important people was coming from Moscow to inspect the club and its proficiency. Their instructor mentioned casually to Valya that one of the members wanted to talk to her.

He turned out to be an official of the All-Union Voluntary Society for the Promotion of the Army, Air Force, and Navy, an organization that promoted better relations between the civil population in the USSR and its armed forces. Valya did not know, of course, that this organization had the task of interviewing and evaluating potential women cosmonauts, all of whom had to hold the rating of parachutist first-class.

She talked for more than an hour with the man from Moscow before it began to dawn on her just what he was after. He was a personal answer to her letter!

He explained to her that she would have to go to Moscow for a series of examinations that would take at least a month. The whole business was very secret. She could not even tell her bosses at Red Perekop or her family. She was given a "cover story" to account for her trip away from Yaroslavl. She was to say that she was taking a medical examination to see if she could qualify for the USSR's National Parachute Team.

Valya spent a month at the Scientific Research Institute of Aviation Medicine secluded in a birch grove in Petrovsky Park on the outskirts of Moscow. This was where Gagarin, Titov, and the other male cosmonauts had been examined. As they had done, she went through all the tests, and she passed them all. Then she returned to Yaroslavl for a week to finish up her business there before moving to Star Town, where the cosmonauts lived and trained. She still could not tell her family that she was now in her country's space program. It would be some time before they found out.

On March 2, 1962, Valya left Yaroslavl, taking her first step on her voyage into space.

A Spaceship for a Sea Gull

The theoretical foundations for manned space flight, and thus the Vostok spaceship in which Valya would orbit the earth, had been laid in nineteenth-century Russia, when Czar Alexander III was on the throne.

They were formulated by a visionary young schoolteacher in Kaluga, a small, rural village some 160 miles south of Moscow. As early as 1883 this remarkable man had worked out the mathematical equations that explained rocket flight. By more than twenty years he preceded the work of Dr. Robert H. Goddard, in Massachusetts, and that of Hermann Oberth, in Romania.

Konstantin Tsiolkovsky was a teen-ager when he first became interested in space travel. Scarlet fever had left him with severely impaired hearing, as it did so many victims in those days before the discovery of antibiotic drugs. This handicap turned young Tsiolkovsky to an introspective and contemplative life. He turned from sports and social functions to books. Another influence also helped to shape his career in science and mathematics. He was fascinated by the science-fiction novels of the Frenchman Jules Verne. In very practical and realistic ways, they described space travel to the moon. Tsiolkovsky thought that with proper scientific knowledge man could bring reality to Verne's fantasies.

At great personal sacrifice, Tsiolkovsky attended lectures in science and mathematics at the university in Moscow that he felt were necessary for developing a theory of manned space flight.

He later became a schoolteacher in order to earn a living, but he never gave up his interest in space flight. After school and during vacations, he would work for hours formulating the mathematical equations that explained how only rockets could escape the tremendous pull of earth's gravity and allow man to reach the moon and the planets. He also theorized that rockets of the future would use liquid oxygen and liquid hydrogen, and this was decades before either gas had even been liquefied. His book *Free Space* appeared in 1883. In it he described how a person in space would react to the strange environment of weightlessness. In 1903 his famous paper "Exploring Space with Reaction Devices" appeared. In it he laid down the theoretical basis for astronautics, including what is known as the "rocket equation" in the United States but is called the "Tsiolkovsky equation" in the USSR. He also proved mathematically that the multistage rocket was the only answer to reaching the moon and the planets.

Despite these accomplishments, Tsiolkovsky died relatively unknown in 1935. He had published a number of scientific articles on aviation and space flight, and he had been made a member of what was later to become the Academy of Sciences of the USSR. However, in the lean years following the Great October Revolution of 1917, there was little money available for scientific research—and none for the visionary in Kaluga whose work was to lead directly to the landing of the first man on the moon.

Tsiolkovsky's work and writings did influence a man of vision and talent who later put Valya into space. They captured the imagination of a man who became "the Von Braun of the Soviet Union." He was Sergei Pavlovich Korolev, long known outside of the USSR only as the Chief Designer of spacecraft.

Korolev was born in the Ukraine, on December 30, 1906. His parents were intelligent, well-educated people who were later divorced. His mother remarried, and Sergei's stepfather, an engineer, moved his new family to Odessa. Since there was a seaplane base in that town, young Sergei soon became fascinated by airplanes. It was not long before he had talked a pilot into taking him on a flight. He became hopelessly addicted. He talked, thought, dreamed, and read about nothing but airplanes.

In 1923 young Korolev joined a club and learned to fly gliders. Later he worked his way through Kiev Polytechnical Institute and the N. E. Bauman Higher Technical School in Moscow, majoring in aeronautics and graduating in 1930. While there, he was a pupil of Andrei Nikolayevich Tupolev, one of the greatest of the Soviet Union's aircraft designers.

In 1931 Korolev was appointed director of the laboratory of the Group for the Study of Reactive Propulsion, in Moscow. Like the Society for Space Research in Germany, the American Interplanetary Society, and the British Interplanetary Society, it was an organization of talented and dedicated men who were absorbed in rocketry and space flight.

In this position, Korolev came to know and admire Fridrikh Arturovich Tsander, another famous name in the history of rocketry, a man whose dedication to space travel led him to name his son Mercury and his daughter Astra. Tsander designed some of the first liquid-propellant rocket engines in the USSR.

On August 11, 1933, Korolev was in charge of the first launching of a liquid-propellant rocket built by the group. It failed to leave the pad, and it was one of the few failures that Korolev was to experience in thirty years.

During World War II, Korolev was engaged largely in designing small rockets to help heavily loaded aircraft take off. However, toward the end of the war, he became involved in the design of military rockets.

Following World War II, when the more affluent nations began building intercontinental missiles, Korolev was appointed chief design engineer for such weapons at Scientific Research Institute No. 88 in Kaliningrad, near Moscow.

When the Space Age dawned, there was obviously only one man in the USSR qualified to head his country's effort: Sergei Pavlovich Korolev.

His first ventures into space were made in adaptations of the German V-2. The Soviet version, called the V-2A, and the later V-5V, were used to launch scientific instruments as well as dogs into the lower reaches of space. However, these payloads were not designed to orbit. By the mid-1950's, these rockets launched from the range at Kapustin Yar on the Volga, the USSR's first spaceport, were sending mongrel dogs named Whitey, Gnat, Little Bit, and Gypsy to altitudes of 150 to 280 miles. The nose cones in which they rode were recovered by parachute.

In 1953 Korolev was given the assignment of developing his country's first intercontinental ballistic missile. He did this using his own design team and manufacturing facilities. Captured German rocket scientists taken into the USSR by force in 1946 and 1947 played only a minor role in his effort. They were used as consultants in specialized areas, such as gyroscope design and high-temperature metallurgy, in which the Russians lacked the required knowledge.

In the mid-1950's Korolev proposed to the Central Committe of the Communist Party that a satellite be put into orbit about the earth. In July and August 1957 the first missiles were tested by launching them from the USSR into the Pacific Ocean. Soon afterward the Central Committee told him to go to work on the satellite.

On October 4, 1957, one of Korolev's modified ICBMs launched the world's first artificial satellite into orbit from Kapustin Yar. It was *Sputnik 1,* a shiny aluminum sphere 22 inches in diameter and weighing 184 pounds! On November 3, while the world was still marveling over *Sputnik 1, Sputnik 2* thundered into space. It contained the first living creature to be sent into orbit around the earth: a mongrel dog named Laika ("Barker") . She orbited at an altitude of between 140 and 1,038 miles and lived for a week. Instruments attached to Laika transmitted signals to earth that proved she could live in weightlessness and was eating food and drinking water.

Korolev then turned his talents to space in earnest. With this new booster, the USSR launched *Luna 1* to the moon on January 2, 1959. It missed its target by some 3,600 miles, but *Luna 2,* launched on September 12, crashed into the moon as planned. On October 4 Korolev supervised the launching of *Luna 3* from Kapustin Yar. It circled behind the moon and took the first pictures of its hidden side.

Now it was time to think about sending man into space. The development of the spaceship that would do this was not a smooth process. Korolev presided as a benevolent, technical dictator, but he was shrewd enough to know that he did not have a corner on all the knowledge in this new area that combined widely different areas of science and engineering. So he consulted frequently with many distinguished Russian biologists and physiologists. They knew a lot about the effects of radiation on men, and he needed their aid to ensure that the spacecraft would provide adequate shielding against the deadly radiation of space.

Korolev also knew that the Americans were committed to a conical shape for their Mercury spacecraft. It had a distinct and highly desirable advantage. Mercury could be maneuvered during reentry through the earth's atmosphere. The astronaut could guide it to his target. However, such a shape required sophisticated and specialized equipment such as high-speed computers and supersonic wind tunnels that simply were not available to the Russian space program.

Korolev knew that he would have to cut corners and compromise somewhere in the interests of time. He was under great pressure from the party to make sure that the Russians were first into space with a man.

The booster-rocket designers had placed some constraints upon his spacecraft designers. The booster could lift only 10,000 pounds into an orbit below the Van Allen radiation belts. Furthermore, the booster capabilities dictated that the spacecraft could be no more than 9.6 feet in width or diameter and no more than 15 feet long. At last he had something concrete upon which to start work.

Recalling the theoretical studies of Tsiolkovsky, he

conceived of a two-part spacecraft. One part would hold the cosmonaut and everything that it took to keep him alive and to provide him with the means of controlling his ship. The other part would contain the expendables—the equipment that would not be needed during the return to earth. This part need not be brought back at all because it was dead weight.

The younger engineers and the older ones from the 1930's disagreed on one feature: the cosmonaut's cabin. A spirited discussion arose between the two factions. Korolev listened dispassionately. He knew that both sides had equally valid and scientifically sound arguments. However, he knew the ultimate decision was his.

The older hands preferred a cautious and conservative approach to design. They pointed out that basically it was merely a problem in ballistics—like throwing a rock into the air, someone said. Thus, the best shape for the reentry body would be a sphere.

The older engineers argued that the sphere could withstand the heat of reentry better than a cone. They added that a sphere gives the maximum internal volume of any shape, and is aerodynamically stable at all velocities.

To prove this point, assistants climbed up a ladder and tossed Ping-Pong balls weighted with a bit of plastic into the air to demonstrate their stability. Later a full-sized sphere would be mounted in an airplane. At 32,000 feet it was pushed out, and demonstrated in full scale what the Ping-Pong balls had proved when thrown from the ladder.

The younger engineers agreed that the sphere was fine but that the Americans had proven the advantage of using a cone. All they needed was time to work out the optimum design . . . .

Time was the one thing Korolev did not have. The party dictated the launch schedules—not science, not technology, not the Chief Designer.

"Now, *everything* considered," he said to bring the argument to a close, "don't you agree that the sphere is the best way to go?"

The loyal opposition, with reluctance, agreed.

Thus was born the spherical Vostok, the offspring of political necessity and technological compromise. It received its name when a young engineer suggested they call it Vostok, "the East."

What was Vostok like? Vostok was a sphere with an offset center of mass. It soon became known as the *sharik* ("ball"), and engineers and cosmonauts alike adopted the word much as their American counterparts used the word "capsule" to refer to their Mercury spacecraft.

In principle, the *sharik* behaved like the "goofy golf ball" that practical jokers substitute for a good one in a game against unsuspecting friends. It is a ball with a weight placed off-center inside it. When putted, the ball wobbles about in a curving path and misses the cup completely. The *sharik* rolled in exactly the same way on the ground after landing, which is one reason why the cosmonauts landed separately by parachute!

However, Vostok was designed for use in space and not for rolling on the ground. When the cosmonaut was ready to return from orbit, he used the attitude-control rockets of the service module to position the Vostok so that it entered the earth's atmosphere at the appropriate angle. Then he fired the retrorocket. Freed from the service module, the *sharik* entered the air with its offset weight pointing to the center of the earth, and the craft described a curving path like that of a rock thrown into the air.

The sphere had its disadvantages, though.

Like the rock, it would land where it would. The cosmonaut could not maneuver the *sharik*. Depending upon the wind and other atmospheric conditions, the *sharik* could and often did spin around like a top. With no attitude-control system, the cosmonaut simply had to sit, grimly thankful that he had been prepared for the ordeal by riding the rotator in Star Town.

Before Korolev sent Yuri Gagarin into orbit on April 12, 1961, he thoroughly tested both the Vostok spacecraft and its space booster. His caution proved to be well taken. The first test flight of the Vostok spacecraft was on May 15, 1960. The mission was known as *Sputnik 4*. The cosmonauts on this mission were a collection of mice, fruit flies, and algae. Four days after launch, the signal was sent to the spacecraft to orient itself for retrorocket firing. Something went wrong. The attitude-control system malfunctioned. When the retrorocket fired, it sent the spacecraft into a higher orbit rather than sending it back to earth.

The test was repeated on August 19 with two dogs named Strelka ("Little Arrow") and Belka ("Squirrel") in *Sputnik 5*. This time everything worked fine. The dogs came down and were ejected from the Vostok at an altitude of about four miles. They were recovered by parachute and were unhurt. Still Korolev was not satisfied. He insisted on another test. Again, it was well that he did. On December 1 another Vostok, called *Sputnik 6,* took off from Tyuratam and entered orbit with two more dogs called Pchelka ("Bee") and Mushka ("Little Fly"). The attitude-control system again did not operate properly. When the retrorocket fired, the angle was too great. The spacecraft plunged to earth as a fiery "shooting star."

Korolev called off everything until a thorough study had been made of the booster rocket and the Vostok.

The next check flight was made on March 9, 1961, in *Sputnik 9*. Again a mongrel dog, called Chernushka ("Blackie"), was the cosmonaut. The mission was a success, and the dog survived to be recovered by parachute. Still, there was a nagging doubt in Korolev's mind. Cautiously he ordered another spacecraft and dog. On March 25 *Sputnik 10* lifted off from Tyuratam. In it were two strange traveling companions. They were Zvezdochka ("Little Star") and Ivan Ivanovich ("John Johnson").

Zvezdochka was a mongrel named by Yuri Gagarin, who was at the spaceport to see her off into orbit. Ivan Ivanovich was a dummy figure rigged up to send tape recordings to check out the spacecraft's communications system and the operation of the ejection seat. "Ivan Ivanovich" was the whimsical name Soviet air force pilots used for the sandbag ballast that trimmed their airplanes when they flew without a co-pilot in a two-seater plane. Even more whimsical was the tape selected for Ivan Ivanovich to transmit—a recording of songs by the famous Pyatnitsky Chorus. Thus Western eavesdroppers who monitored the mission might presume that the Soviets had orbited a concert choir!

With the successful conclusion of the *Sputnik 10* mission, Korolev gave permission for the first manned Vostok.

On April 12, 1961, *Vostok 1* with Yuri Gagarin raced around the earth in 89.1 minutes, with a perigee, or closest point to the earth, of 112 miles and an apogee, or point farthest from the earth, of 203 miles.

Sergei Pavlovich Korolev had proved that imagination, inspiration, and intuition, combined with dedicated and uncompromising discipline in engineering, an absolute faith in man's destiny in space, and the resources of a technologically advanced nation could send man on his first steps to the stars.

Yuri Gagarin proved that the *sharik* could take it.

Vostok looked like a huge silver golf ball. It was 7.5 feet in diameter and was attached to a conical service module by four metal bands that on the detonation of explosive bolts separated it from the *sharik*. The service module served as a convenient place to mount the spherical bottles containing nitrogen and oxygen for the cabin atmosphere, and the radiator for controlling the temperature within it. It also provided space to mount batteries and electrical components and some of the radio antennas. The structure also housed the retrorocket for braking the Vostok out of orbit and the attitude-control system.

All in all, Vostok had some 300 systems and subsystems, consisting of 240 vacuum tubes, 6,300 transistors and diodes, 760 switches and relays, and 9 miles of wiring!

The *sharik* had a rigid metal structure some two inches thick. The metal hull was covered with alternate layers of ablative material in honeycomb cells. This material consisted of fiberglas layers embedded in a resin-epoxy base. Over this was placed a layer of aluminum foil. As the temperatures of reentry built up, this material melted and turned to a gas that transferred heat to the surrounding atmosphere, thus protecting the innermost shell of the *sharik* from melting. There were only six openings in the basic structure: three large holes and three small ones, the latter being windows. The largest hole was the hatch by which the cosmonaut entered and left the *sharik*. Another hatch covered the recovery-parachute container, and a third one was a connector for electrical fittings on a cable arm from the service module. The three windows were covered with high-temperature-resistant glass that would not melt during reentry. As added protection, they also had shutters to keep out the heat and blinding sunlight.

Inside, the *sharik* bore little relation to the high-speed jet aircraft to which Gagarin, Titov, Nikolayev, and the other four cosmonauts of the first class were accustomed.

Sitting in the ejection seat in the center of the *sharik*, the cosmonaut had between his feet one of the three windows. It pointed directly down to earth and was covered with a special grid called the VZOR, which permitted the cosmonaut to orient the Vostok visually. To his left and right at head level were the other two windows. Also to his left was the "trunk," or compartment in the wall that held the parachute. On the same wall was located the water tank, with a ten-day supply of water especially blended for his personal taste and protected by chemicals from the growth of bacteria. A plastic tube with a nipple valve on it permitted the cosmonaut to have a drink, provided the faceplate on his helmet was not closed!

On the left side of the *sharik* also was a panel with an emergency temperature control system, the beacon transmitter for homing signals after landing, and a tape recorder that automatically recorded every word spoken by the cosmonaut during his mission.

Above the VZOR window was the cosmonaut's control panel. On it were the meters that displayed vital information about the *sharik*'s internal environment. There were dials that displayed temperature, pressure, relative humidity, oxygen content, carbon dioxide content, and pressures in other critical areas of the spacecraft, as well as an elapsed-time clock started at the moment of lift-off. Also on this panel was a small globe like the one used in teaching geography. Over it was a grid of fine lines. The globe turned continuously as the mission progressed, and it was synchronized with a much larger one in the mission-control system at Star Town. Thus, the cosmonaut and Sergei Pavlovich knew at any moment what spot on earth the Vostok was over. The device also was used in helping the cosmonaut determine where he would come down if he had to make an emergency landing on earth.

Just below the control panel was a television camera focused on the cosmonaut's face. Another was attached on the right wall to provide a side view of him.

Within easy reach of the cosmonaut's right hand were the *sharik*'s food compartment, controls for the air-conditioning system, some of the radio equipment, an electric clock, the sanitation system for the cosmonaut, part of the electric power supply, and the hand control for manually changing the attitude of the Vostok.

In compartments beneath his feet and behind the ejection seat were the telemetering equipment and the cosmonaut's life-support system. The former radioed important physiological data from the cosmonaut and environmental and functional data from the *sharik* to the mission-control center. The life-support system automatically kept the temperature between 54° F. and 77° F. at a pressure of 14.7 pounds per square inch. It also maintained a mixture of 24 per cent oxygen and 75 per cent nitrogen and kept the carbon dioxide level to less than one per cent. The humidity could be controlled between 30 per cent and 70 per cent for comfort.

The Vostok always carried food, water, air, and electric power for ten days in space—no matter how long the mission was to last.

The ten-day reserve was chosen for a special reason. If the retrorocket failed to fire and brake the Vostok out of orbit when the mission was completed, the reserve would keep the cosmonaut alive until the satellite decayed naturally from orbit. Even at the altitude at which the Vostok circled there were enough air molecules to slow the ship down gradually. Over a ten-day period, it would spiral slowly around the earth through the atmosphere and land on the earth.

The ejection seat for the cosmonaut was a masterpiece of engineering design. Korolev had foreseen that it would have to do much more than be a chair and a bed. It had to immobilize the cosmonaut so that vibration, acceleration, and deceleration forces would be minimized. It also had to be relatively comfortable. Most importantly, however, it had to provide him with a means of escaping from the *sharik* in case of an emergency on the launching pad and serve as a routine means of leaving the spacecraft after reentry into the atmosphere.

The seat was an adaptation of the one used in the Mig-19 jet fighter. Basically it was a padded aluminum chair with wheels and mounted on two rails. At its base were two small solid-propellant rockets. On ignition these sent the seat and cosmonaut rapidly up the short tracks and through the open hatch of the *sharik*. The hatch was blown off by explosive bolts two seconds before the rockets ignited.

The seat was also equipped with a parachute and a survival pack with an emergency supply of food and water, an inflatable rubber raft, and a rescue radio beacon which search aircraft and boats could "home in on" in case the cosmonaut came down in unplanned areas of the earth. This package dangled at the end of a 48-foot-long line beneath the cosmonaut.

If the cosmonaut elected to ride the *sharik* down and not bail out, which only Yuri Gagarin did, at the cost of a very rough landing and some bad cuts and bruises, the rockets beneath the seat were not fired. Instead, at an altitude of about 9,500 feet the hatch on the parachute compartment would be blown off by its explosive bolts. A small drogue parachute would be deployed that pulled out the main chute when the *sharik* reached an altitude of some 7,700 feet.

From this description, it is apparent that Korolev had done a very good job of engineering design.

When he invited the first detachment of cosmonauts in 1960 to visit the plant where the Vostok was assembled, he was proud of his design. However, since he was a shrewd engineer and pilot, he knew the value of consulting the user of any product, especially in aircraft and spacecraft.

"Now, I want you fellows to let me know what you think about the design. Don't be afraid to speak up. Let me know what you think," he said with emphasis.

The cosmonauts looked at drawings and specifications. Then they climbed all over the *sharik*. They sat in the seat and flipped switches and reached for the water hose, the handle on the door of the food locker, and the relief tube of the sanitation system.

Sergei Pavlovich was especially interested to know what Pavel Belyayev and Vladimir Komarov thought. These were men with much experience in flying high-performance jets, and they also had advanced engineering degrees in aeronautical sciences. He was also curious as to what "Shorty" Gagarin would say. Here was a hot pilot, short on classroom training and degrees but not lacking in courage, initiative, and lots of jet time.

In the responses, Sergei Pavlovich was disappointed. The fledgling cosmonauts were too much in awe of his engineering brilliance to dare make suggestions.

Except, of course, for the one guy that would be expected to sound off: Gherman Titov, who would have been the first man in space if politics had not intervened.

Titov was not awed by authority. He had only missed being the first man to orbit earth because he was just a little too brash and self-confident and did not have the proper background from a propaganda standpoint for the Party. Gherman, with characteristic initiative, took Sergei Pavlovich at his word. He went back to Star Town and made several drawings of changes he thought would benefit the Vostok.

The Chief Designer looked them over. At first he was angered by the audacity of this jet jockey. But he studied the sketches and notes that Gherman submitted. They had merit, genuine merit. Gherman had backed up his suggestions with sound engineering knowledge. That was enough for Sergei Pavlovich. The engineering-change order was issued.

The next time the cosmonauts visited the plant, Sergei Pavlovich singled out Titov. Gherman lowered himself into the *sharik* and glanced around. He reached forward and noted that several switches were much more readily attainable. He also noted that the numbers on several meters were larger and much easier to read in the dim light. The major change, however, was in the seat. It was now mounted on a gimbal so that it would rotate from side to side. The cosmonaut could turn it easily to see out of the windows. All his suggestions had been taken!

"Don't get any more ideas, Titov," Korolev warned. "Don't think I am going to completely rebuild the Vostok just to suit you cosmonauts."

While design and fabrication of the Vostok was going on, its space booster was in its last stages of development and testing.

Korolev and his long-time friend and co-worker Valentin Glushko felt more at ease with the booster than the spacecraft. After all, both had been working on its basic building block since 1953—and they had been building rockets for thirty years.

The booster that launched the first probes to the moon and placed the first men and women into space was a modification of the Soviet Union's first large ballistic missile. The huge rocket was ninety feet tall and twenty-five feet in diameter at the bottom. Fully loaded with liquid oxygen and kerosene propellants, it weighed 600,000 pounds. It was a multistage rocket with thirty-two engines all burning together at lift-off. They produced about one million pounds of thrust, or some one-seventh of that created by the Saturn 5 rocket that launched the first astronauts to the moon.

This was to be Valya's booster, and a big one it was.

Sea Gull in Star Town

About forty-five miles northeast of Moscow is Zvezdny Gorodok ("Star Town"), the home and training center for the Russian cosmonauts. In many ways it can be compared to the Johnson Space Center, in Houston, Texas, where American astronauts train. When Valya arrived there for the first time, Star Town was only two years old.

Before leaving Moscow for Star Town, Valya had an appointment with Lieutenant General Nikolai P. Kamanin, an air force officer who had been in charge of planning Star Town and was now its commander. Valya reported to his office at precisely 10:00 A.M., as the telegram had ordered her to do.

With General Nikolai Kamanin, commander of the training center at Star Town.

As she walked into his office, General Kamanin got up from his desk and shook hands with her. "You, Valentina Tereshkova," he said, "are the first and, for the time being, the only woman cosmonaut. But don't worry—there will be others."

He told her that many women had applied but only four were chosen for the first batch. She had been the first selected. The others would arrive at Star Town within the next few days. "However," he added with a grin, "you weren't the first to apply. Do you know who she was and when she applied for cosmonaut selection?"

Valya told him she could not imagine who it was.

"In 1927," he said, "the German Max Valier announced that he was preparing to fly to the moon."

Valya had never heard of Valier.

"A young Soviet woman named Olga Vinnitskaya heard about it and wrote to Konstantin Tsiolkovsky for his advice," the general continued. "She wanted him to intercede for her with Valier. She wanted to go along with him to the moon." The general laughed. "Tsiolkovsky wrote to her that she was being a little premature," he went on. "He said that the newspapers always tended to exaggerate."

Just then a young air force medical officer entered the office.

"Ah, you are ready for your new patient?" asked General Kamanin.

"Yes, General, if you are finished with her," the doctor replied.

However, it turned out that Valya was not really his patient, only his charge. He was to take her to her new home in Star Town.

The first stop on the way to Star Town was the railway station, to pick up Valya's trunk. Having put it in the car, they left the city and headed northeast. Valya was surprised at how quickly they seemed to be in the country. They sped through small villages of log huts that reminded her of Maslennikovo. Some huts were painted bright green with blue trim, and each had a fence around its small garden. Women who could have been Valya's mother or grandmother were pumping water from wells. However, there was one difference. Here and there she could see a television antenna. The resemblance to Maslennikovo ended on that note of technological progress.

After half an hour, they turned off the main highway and onto a secondary road that ran through a forest of birch, pine, and fir.

The car stopped at a tall green fence, and a guard checked the papers that the young doctor handed him. He nodded and then waved his hand for the driver to enter. Valya noted that there was no sign at the gate to indicate what the place was.

Patches of March snow were still all over the ground, and the car splashed in and out of icy puddles.

Dressed in a green overcoat and wearing a golden kerchief, Valya squished into the lobby of the dormitory in her thoroughly soaked low-heeled shoes, hardly the picture one would expect of the first woman cosmonaut.

It was Saturday, and things were dull. Valya expected to see Gagarin and Titov, but they were not there. However, several of the other cosmonauts were around. Aleksei Leonov was sketching, as he always was when he was not training. Pavel Popovich and Valery Bykovsky were playing billiards. Others were watching television. With them was a short, quiet, darkly handsome young man.

"This is Andrian Nikolayev, Valya," someone said.

Valya stuck out her hand and smiled. Andrian took it shyly but could not bring himself to smile back.

"Here, we'll give you a hand with your bags," said Pavel Popovich. He and Andrian grabbed them up and went up the stairs to the second floor.

"We've given you what we call the bridal suite," Pavel said as he opened the door. "It's the sunniest room in the building."

As usual Pavel had his military cap cocked back on his head in a manner absolutely forbidden by regulations. It was doubtful that even General Kamanin could have gotten him to wear it properly. It was his trademark.

They took Valya's bags into the room. Andrian had been carrying the trunk. He put it down, gave the room a glance with his dark, "velvety" eyes (as Valya was later to recall them), and left quietly.

The room later did turn out to be the bridal suite, and

the future bride's future husband had just set down a heavy trunk in it. But at the time Valya was far too absorbed in Star Town, her fellow cosmonauts, her new room, to think about anything else—except, perhaps, Andrian's dark eyes.

The view through the window was fabulous! Not too far away were the pines and firs that screened Star Town. Valya had arrived rather late in the afternoon, and the sun was beginning to set behind the trees. Flashes of gold and red shot among the trees and into the room.

Andrian, Valya would later discover, was different from the rest of his fellow cosmonauts. Andrian was born in a small village on the banks of the Volga in the Chuvash region of the USSR. Before becoming a cosmonaut, he had been a fighter pilot. Before that he had served as a radio operator and machine gunner in a bomber crew. As a teenager he had been a lumberjack in Karelia, in the far north of the USSR.

The sun had set among the pines, and Valya had missed the call to supper. She had to dine alone. The cosmonauts certainly dined in style! She had a fresh cucumber salad in March. And there was a bowl of chicken broth, beef stroganoff, and a glass of cherry juice.

While Valya was marveling over the meal, the young doctor who had accompanied her to Star Town came in. He asked how she was doing, and then he revealed something else. "Valya, day after tomorrow, on Monday, the rest of the women cosmonauts will be here," he said.

Instantly, Valya felt less lonely. She still was uneasy among so many men. It would be nice to have some young women around.

"In the meantime, Valya, make yourself at home, and get used to the routine here," he added.

Since television was something that Valya had never known, she spent the evening in front of the set. While she watched, she happened to sneeze once or twice. Nina, the woman who had been appointed to look after her and the other female cosmonauts, immediately called a doctor. He popped her into bed with several aspirin. She slept soundly all night long.

"Valya, Valya, wake up—it's breakfast time!" shouted voices through her door.

It was the same old story. Her first day on the job and Valya was late for work.

She scrambled out of bed and was faced with her first decision as the first female cosmonaut in the history of the world. What should she wear to breakfast?

She picked out a simple dress and combed her short blond hair. Then her face lit up in the grin that was to become famous all over the world only fifteen months later. As she walked into the dining room, she could tell from the reaction of her fellow cosmonauts that her choice of dresses was the right one.

The cosmonauts all greeted her cordially, but they had something more important than her dress on their minds. Even though it was Sunday, they were on their way to an examination in advanced mathematics. The last few of them took a second cup of coffee, and with quick smiles and waves to Valya headed toward the classroom building.

What did the world's first woman cosmonaut do on her first day in Star Town? She wandered around, getting her feet wet again and sniffling. She watched more television. She went back to her room to rest—and found a bouquet of flowers on her table! She never found out who put them there but wondered if it might not be one of the cosmonauts. By that night she was desperate for the companionship of girl talk. For what seemed like the hundredth time, she picked up the latest copy of *Neptune*, the newspaper put out by the cosmonauts. She smiled again at the funny stories by Titov and Popovich and admired the sketches by Leonov.

Monday, the other young women arrived. Now there was too much girl talk. It seemed to Valya like a gathering of magpies. In less than a half hour, one thing had been established. They were all parachutists first-class, and they had all been interviewed by a man from Moscow who was with the All-Union Voluntary Society for the Promotion of the Army, Air Force, and Navy.

Valya was at once at ease again. Here were young women her own age. One of them was a pilot who knew how to fly jets. True, they were all cosmonaut candidates, but could she compete with them? Could a textile technician compete with candidates who had been to college and could fly airplanes? She looked them over coolly, professionally, but with no jealousy. Valya was self-confident enough to know that she could hold her own in any competition, especially parachute jumping.

Before the training program began, Yuri Gagarin, as commander of the cosmonaut detachment, took the young women on a tour of Star Town. One of the sights they had already seen. On either side of the main road from the gate to the yellow plaster headquarters building were gleaming metal frames. Two pictures were already displayed: those of Yuri Gagarin and Gherman Titov.

"There's room for many more, if you will notice"—Yuri grinned—"including some of you girls."

The training buildings and the gymnasium brought a mixture of reactions. The trainees were most impressed by the "devil's merry-go-round," or centrifuge, and the "iron maiden," or rotator, both of which they were to become very familiar with, and very soon.

Yuri still was a little stiff and formal, as might be expected. During the tour of the museum, which was just getting started at Star Town, he sounded almost like a high-school teacher. He pointed to models of the unmanned satellites and space probes and described them in dry, technical terms. Some of his audience stifled yawns.

During the tour of Gagarin's office, in which he showed Valya and the others some of the many souvenirs he had collected as a result of his flight, someone asked who would be the first man on the moon.

In replying, Gagarin made one of the few inaccurate statements he was ever to give them. "No one yet knows his name," he said, "but you can be assured he is studying, working, and living among us here."

Things became more lively when they met the wives and children of the cosmonauts, most of whom were married.

Valya was especially impressed by the dark beauty of Yuri Gagarin's wife, Valya, who worked as a laboratory technician in the hospital. And Gherman Titov's Tamara was equally beautiful. Vladimir Komarov's wife was named Valya, too! She had studied history in college and was librarian at Star Town.

On Tuesday, formal training started for Valya and her colleagues.

Despite the fact that they had all had some sort of training in either science or technology, they were overwhelmed at the subjects they found themselves enrolled in. There was astronomy, space biology and medicine, geophysics, astronautical engineering, radio communications, and celestial mechanics. Later there were courses in rocket propulsion. For the first few weeks Valya spent every night poring over her new textbooks to stay up with the others. However, Pavel Belyayev and Vladimir Komarov, who had more technical and scientific training than the other cosmonauts, were always eager to help them.

The training program was not all classwork. The female cosmonauts had to take the same rugged physical education program the young men did. A professional gym teacher put them through a series of exercises on the parallel bars, vaulting horse, and trampoline that would have done justice to Olympic athletes. But none of the trainees could match Aleksei Leonov's records. In one year he bicycled 620 miles, made 200 cross-country races, and skied 186 miles.

In addition to gymnastics, there were organized group sports such as basketball, volleyball, and hockey. The latter sport was restricted to the winter, however, because there was a natural rink. Under the spring sun, the hockey rink became a swimming pool! Since most of the young women were unfamiliar with hockey, Gherman Titov, with characteristic self-assurance and leadership, appointed himself their coach. He showed them how to hold the stick properly and particularly how to defend the goal.

Once they acquired a little skill in the game, it was only natural that the tomboy from Yaroslavl should suggest that the girls play the boys.

It really was not much fun—and not much of a contest, either. The girls were playing as hard as they could, but it was obvious the boys were holding back on their skills. They missed the puck with almost comic gestures. Still it was new, and in some ways the girls were sad to see the spring sun put an end to the hockey rink just as they were beginning to learn how to play.

As if these sports were not enough to keep her in shape, Valya also skied, rowed, and raced bicycles. With practice, she cut her time from 3.4 minutes to 3.2 minutes in the 720-meter swim event. By the end of the first phase of physical training, she was able to do twelve pushups on the parallel bars rather than the four she could do when she first started.

Hockey and volleyball, astronomy and physics, were one thing. The "devil's merry-go-round" was something else.

The young men were used to the forces of acceleration. They were all jet pilots. They knew what to expect when they put their plane into a tight turn or pulled out of a dive while flying at high speed. They would be pressed back into their seats, and their weight would increase by a factor of six or eight. Thus, a 150-pound pilot "pulling 6 g's" would in effect weigh 900 pounds. His blood would become as heavy as mercury.

The young women, however, were completely unfamiliar with acceleration, except for their jet pilot. It is true that they had heard many lectures on the effects of acceleration on the human body. And they had seen many films shown by the medical instructors. Theory was all well and good, but they had to learn first-hand about the acceleration they would experience when their rocket lifted off. That is where the "devil's merry-go-round" came in.

The huge Swedish-built centrifuge used to simulate the acceleration of rocket launch was located in a low building set apart from the rest of the facilities at Star Town. Pivoted on the floor and free to rotate about its center was a lever arm fifty feet long. At the end of it was a seat from the Vostok spacecraft. It was arranged in a set of gimbals so that the acceleration force could be directed through the cosmonaut's body as it would be on lift-off. In addition, the speed of the centrifuge could be changed to simulate the accelerations given to the spacecraft by the two stages of the booster.

The first time Valya and the other young women saw the centrifuge, Andrian Nikolayev was riding it. They could see his face on the screen of a television monitor. They were horrified! The dark, handsome face with the thick eyebrows was contorted like that of a monster. Andri-

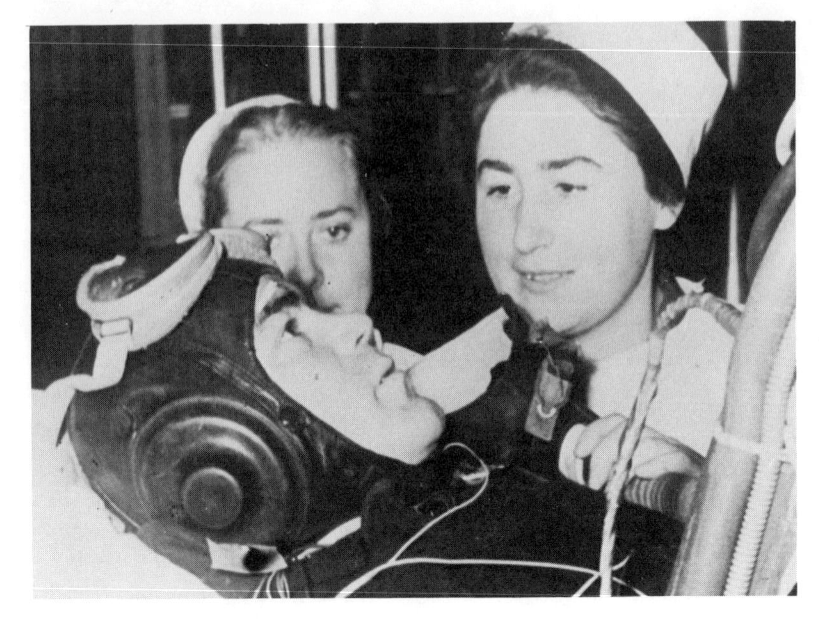

Andrian Nikolayev in the "devil's merry-go-round."

an's lips were peeled back in a hideous grin, and his eyes sank back into his head, the lids unable to close.

The instructor in charge of the "devil's merry-go-round" asked who was going to be first.

There was not an immediate rush to get into the seat. But Valya, who could never turn down a challenge, stepped forward. The instructor made a mental note that she was the first to volunteer.

Strapped into the seat, Valya made her first revolution. She grinned broadly, and suddenly she could not stop. Her lips drew back just as hideously as had Andrian's. Her body felt as though she were being crushed, and her breathing became very difficult. Then she noticed what some of the cosmonauts had told her. On the centrifuge your blood becomes very heavy. Valya became aware that her heart had to pump furiously to move it.

Suddenly the pain began to lessen. The machine coasted to a stop.

"Well, how was it?" asked the instructor.

"It wasn't too bad," Valya lied.

Back in the dormitory, the others clustered around Valya. They wanted to know what it really was like.

"It was horrible," said Valya—and let the matter drop.

Valya felt depressed after the ride, but finally she realized what was worrying her. The reason she had been in such great pain on the machine was because she was terrified. It was all mental—not physical. The machine was only pulling 3 g's. The young men regularly took loads as great as 6 to 8 g's. She simply did not know whether she could hold up under a stress like that. In short, Valya had experienced the same sort of tension she always felt before a parachute contest, especially when she was competing with young men.

If the centrifuge was the "devil's merry-go-round," then the "iron maiden" must have been his food blender. It was a fiendish device designed to train the cosmonauts in what it feels like to be in a spacecraft out of control.

It was a metal box, "just about the size of a coffin," said Gherman Titov with his ghoulish sense of humor. The "iron maiden" was suspended by three gimbals, like a gyroscope, so that it could tumble on all three of its axes at once. The trainee sat strapped in a Vostok spacecraft seat, surrounded by foam padding that cuddled around his body, "in a horrible, gooey mess like gelatin," according to Gherman. With the lid closed and bolted, the would-be cosmonaut was in total darkness. As the machine gathered speed, tumbling madly end over end and every other way at the same time, doctors and psychologists kept up a constant flow of questions to the befuddled occupant, who was expected to answer all of them logically and coherently.

The other training machines were less terrifying, but it took plenty of hard work to keep up with them.

The swinging table was no worse than being on a ship in rough seas. Valya had to keep her balance in the middle of the table or walk about on it as it was suddenly tilted. The vibration machine had her sitting in a chair that bounced at the same frequencies she would experience when her booster rocket lifted off from the pad.

Not so easy to endure were two special chambers in which she had to sit from hours on end to days on end.

The temperature chamber could reproduce a wide range of temperatures that Valya had never known in Yaroslavl. It could plunge to the frigid cold of the Arctic at near 0° F. or suddenly rise to the blistering heat of the Sahara Desert at 130° F. While she could adjust her clothing to meet the changes, a day in the temperature chamber was not considered by Valya and her friends as anything but sheer torture.

The isolation chamber was, if anything, worse than the temperature chamber. The air in it was just right, and so was the humidity. There was a cosmonaut's seat and a worktable. Near the table was a console, or electric display panel, with switches and meters that Valya had to operate. There was a supply of space food and water, too. The walls were covered with material that deadened all sound from the outside. In fact, the room was so well insulated that Valya could hear her own heart beating.

She knew, however, that she was not completely alone. There were television cameras that kept her within view twenty-four hours a day.

When Valya went into the chamber, she never knew how long she would be there. It might be only a few hours, several days, or even as long as two weeks. The young men told her how they managed to pass the time in the chamber.

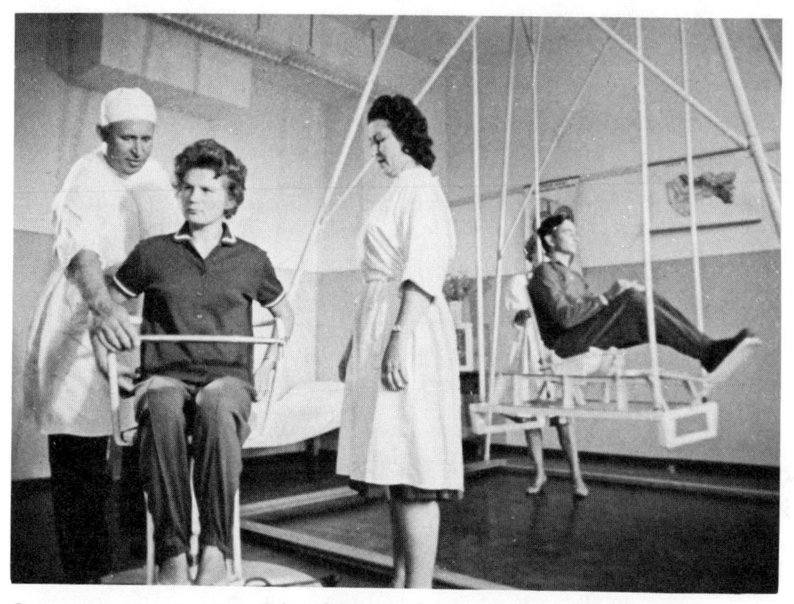

Cosmonauts are tested in this revolving chair and special swing, which Valya and her fellow cosmonaut Valery Bykovsky demonstrated in the Russian film Road to the Stars.

Yuri Gagarin said that he imagined he was already in his spacecraft. He would close his eyes and picture himself orbiting over London, Rome, Paris—all cities he had never seen on earth. Gherman Titov spent his time memorizing whole chapters from his favorite novel. Pavel Popovich, characteristically, sang from his vast storehouse of songs, even making some up when he ran out. The "velvety-eyed" Andrian Nikolayev, as one would expect, solved advanced mathematical problems!

Valya had read many articles about experiments made in isolation chambers. Subjects in Canada and the United States had all but gone mad. They began to see fires or imagine that there were holes in the walls. Valya seriously began to doubt whether she could endure the isolation chamber, despite all the pats on the back from Yuri and Pavel and the shy smile and reassuring words from Andrian. She knew that she was naturally friendly and talkative; she liked to be around people. But there were times when she was also moody. Would the isolation chamber be the one thing that caused her to be thrown out of the cosmonaut program? It was one of the courses that was graded very strictly. A bad mark on the isolation chamber, and Valya would be out of the program.

The medical specialists prepared her for the chamber. First they applied electrical leads to her chest so that they could record her heart activity on an electrocardiograph. Then they attached similar leads to other instruments that made a continuous record of her temperature, blood pressure, and breathing rate.

At first, sleeping was a problem. There was no bed, only the cosmonaut's seat. Valya had to learn to sleep sitting up.

A day passed and then another. She performed her required tasks and kept a record of everything she did. Soon things became a routine, but she constantly yearned for some sort of contact with the outside world. She would have given anything to have a radio or television.

She even thought how wonderful it would be to hear a mouse scurrying through the walls or a cricket chirping. In her imagination, she began to recall all the sounds that were familiar to her. There was the factory whistle at Red Perekop, and the train whistle, too. The sounds of the Volga and the boats on it, the wind murmuring through the forests around Maslennikovo. These memories helped her pass the time.

When Valya left the chamber after a week, the noise of the world almost deafened her. A door shut softly sounded like a cannon going off. Even the sound of her own footsteps was deafening. It took her almost two days to get accustomed to the ordinary sounds around her.

One phase of cosmonaut training presented no problems for Valya. It was parachute jumping.

Lieutenant Colonel Nikolai Konstantinovich Nikitin was a rather short, thickset man with blond hair and a face full of freckles. He was also a holder of several world records and probably the best parachutist in the USSR.

"Well, dolls, let's have a go at it," he flippantly greeted the young women.

Even though he knew that everyone in the group was a parachutist first-class, Nikitin began at the beginning, just as Valya's instructor had done only a few years earlier in Yaroslavl.

He also questioned them about their experience and their instructors.

"Let's see, Valya," he said, "you are from Yaroslavl. That means you were one of Stanislav Morozychev's pupils."

"That's right! Do you know him?" she replied.

"Yes, he is an excellent instructor," Nikitin answered. "He's one of the best in the business."

Valya agreed with him completely.

"By the way, Valya, how are you at delayed jumping?" he went on.

Valya wondered why he would ask such a question unless he knew how much trouble she had with it.

She said rather evasively that it was difficult but she could do it.

Nikitin smiled strangely and went on to another topic.

Valya wondered if she had given the right answer.

She must have, because he never came back to the subject again.

It was not long before the girls, Valya included, developed a crush on Nikitin. It was a peculiar kind of crush, though. He became half father and half boyfriend. When anything went wrong, they ran to him.

After one session on the "devil's merry-go-round" in which she did everything wrong, Valya got what amounted to a C-minus. She went to Nikitin in tears.

"Valya," he said sternly, "you are no longer a kid. You have got to get over the idea that you have to be first in everything."

His psychology worked. It made Valya so mad that she could not wait to get back on the centrifuge and show them what she could really do. And she did!

The parachuting exercises became more and more difficult, but they were no problem for her. She had done it all before. There were night jumps. Also there were jumps by day and by night when she wore her space suit and landed in water. They all required skill and self-assurance, and Valya had enough of each to master the tests put to her. By the end of her training program, she had made sixty-two jumps.

Jumping was not without its hazards. One day while practicing with Valery Bykovsky (who would precede her into space on *Vostok 5*), she suffered an accident. As Valery cleared the door of the airplane, Valya followed him a little too quickly. His boot struck her in the forehead.

There was other training in airplanes as well.

It was in a Tu-104A plane that Valya first experienced the sensation of weightlessness that she would later know for three days. This time her acquaintance with it lasted only forty-five seconds.

Valya and the other young women followed the pilots into the Tu-104A plane and sat down, strapping themselves in. Their instructor explained how the training exercise would work. He told them the plane would nose down a bit and then pull up at an angle of forty-five degrees. It would then fly in a curve like a rock thrown into the air. For about three-fourths of a minute the force of gravity would be canceled by the centrifugal force of the moving plane.

"Only three-fourths of a minute?" someone asked.

"Well, if that's not long enough," he replied with a straight face, "I'll get the pilot to make it forty-five seconds."

He then explained in detail the physical principles involved in the flight maneuver, pointing out that he was just reviewing what the group should already know.

The simulation of weightlessness was a serious test. It may have been fun, but it was not meant to be, as one of the young women found out. Other technicians in the plane watched each student closely. They were making mental notes. The sign on the ceiling flashed on "Attention." The young women unbuckled their seat belts. The sign winked out. Another lit up: "Weightlessness." It was not needed. Some passengers drifted lazily out of their seats and floated toward the padded top of the plane. Others were wafted sideways, bouncing gently off the padded walls. Their reactions to this first experience with weightlessness fell into three categories, as the instructors knew they would. Some would have no problems. They would tumble end over end and, in general, think it was great fun. Others would become suddenly confused and thrash their arms and legs around. Still others would become sick.

This crucial test separated the trainees into three groups. Those in group one had the best chance of going into space. Those in group two would be considered for space flight. Those in group three had no chance.

Valya had no problems, and was placed in group one.

The flights continued for those in groups one and two. Each one was a working session, however. It was not meant to be fun and games. As soon as the "Weightlessness" light went on, Valya would float out of her seat with her space food in one hand and her drinking fountain in the other. The space food was in an aluminum tube, just like toothpaste. In fact, it had the same consistency as toothpaste! The tube had a plastic nozzle on the end, and she placed it into her mouth and squeezed out the chopped meat. It tasted just like baby food.

Drinking water while one is weightless is tricky—and potentially dangerous. The plastic squeeze-bottle has a hose, and a nipple which must be deftly inserted into the mouth. Then the bottle must be squeezed. Water escaping prematurely from such a bottle floats around the aircraft or spacecraft in small, spherical drops that can be inhaled if one is not careful. Likewise, there is a danger from crumbs of food that break loose from space food. These, too, can float into the nose and lungs and cause sneezing or even strangulation.

Before her training was over, Valya became even more familiar with airplanes. She had to learn to fly the Yak-18 trainer. She also flew as a passenger with Yuri and the other pilot-cosmonauts in the Mig-15 jet.

As her training progressed, Valya became more acquainted with the Vostok spacecraft and the simulator, which was an exact duplicate of it. Before her own space mission, she spent many more hours in the Vostok space simulator than she did in the real thing. One of the high points of her training came during an inspection trip to the factory that manufactured the Vostok. All the young women went along, curious to meet for the first time Sergei Pavlovich Korolev, the Chief Designer of Spacecraft, about whom they had heard so much.

They simply did not know what to expect after everything that they had heard. But they did know that Korolev was a brilliant engineer with long experience in space research and rocketry; he was the most influential man in the Soviet space program.

"I ask for cosmonauts, and they send me a bunch of girl parachutists," grumbled a stocky, gray-haired, middle-aged man wearing a short-sleeved shirt and baggy trousers.

The trainees stiffened. Korolev set off at once down a corridor leading to the main assembly room.

Scattered throughout the large room were several spacecraft in various stages of assembly. Just by glancing around, it was possible to get a good overall picture of how complicated the Vostok was. Valya and the others were impressed by the fact that Sergei Pavlovich was constantly being interrupted during their tour by engineers who rushed up to him waving drawings and asking advice. With a quick glance at a drawing or a page or two of a report, he would give a clear and incisive answer, and the engineers would disappear.

As the tour went on, Valya gradually saw that gruffness was merely a facade for Korolev's basic shyness and quiet sense of humor. She and her colleagues soon became accustomed to the irony and gentle wit that were the trademark of one of the world's greatest pioneers in space travel. There were literally hundreds of questions they wanted to ask about the Vostok and its launching vehicle. However, they could see that Korolev had enough on his mind without listening to their amateurish questions, and they kept silent. They knew that he usually put in a twelve- or fourteen-hour day, so they did not chatter as they usually did at Star Town.

In March 1962 the pace of activity at Star Town began to quicken for two of its members.

Valya began to notice that they saw less and less of Andrian Nikolayev and Pavel Popovich. When she asked

Yuri what was going on, he smiled slyly and said that they had gone fishing.

Everyone in the cosmonaut training detachment knew that Andrian was a passionate fisherman and an expert cook of fish as well. His fish soup was far better than they got in the cafeteria! However, happy-go-lucky Pavel was certainly not the type to spend any time fishing. This restless descendant of the Dnieper cossacks preferred dancing and singing to sitting quietly on a riverbank.

It soon became apparent that Andrian and Pavel were preparing for the next two manned spaceflight missions, *Vostok 3* and *4*.

Andrian had been the backup pilot to Gherman Titov, who had piloted *Vostok 2* for seventeen orbits of the earth only a year earlier. And, Valya recalled, it was these same two who had carried her suitcases and trunk up to her room when she first arrived in Star Town. While she missed Pavel's good-natured horseplay and his singing, she also missed Andrian's shy smile and "dark, velvety eyes." The children of Star Town missed Andrian, too. Being a bachelor, his room was almost a clubhouse for them. He was good at settling disputes on everything from the rules of soccer to the ownership of property after several involved transactions of barter or trading. Andrian Nikolayev was also acknowledged to be the best ice-skate sharpener in Star Town.

Valya and the other young women wanted to accompany their fellow students who were going to the launching site with Andrian and Pavel and the backup cosmonauts, Valery Bykovsky and Vladimir Komarov. However, their training schedule would not permit such a trip.

As it was, there was little studying between August 11 and 14. They spent a lot of time listening to the reports coming over the intercom from the launching site. On

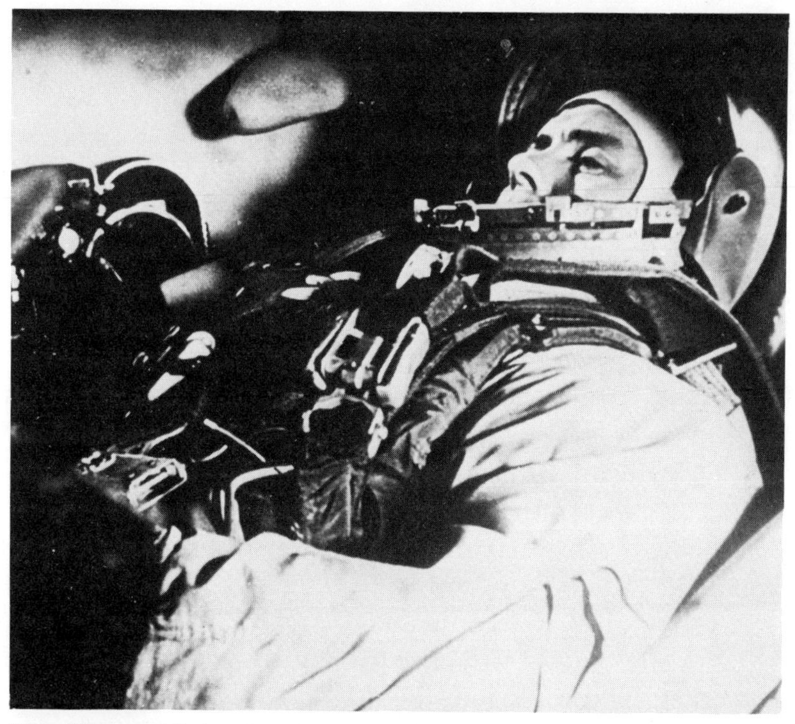

Pavel Popovich in Vostok 4.

August 11 they held their breath as they heard the words "Lift-off! We have lift-off!"

A day later, on August 12, while Andrian was in his eighteenth orbit, Pavel was launched. The young women continued to follow the dual mission, sometimes by radio and sometimes by television, in the mission-control center in Star Town. They stayed with it until August 14, when both men returned to earth via parachute, landing in the hilly country some 1,500 miles southeast of Moscow.

When Andrian and Pavel returned to Star Town, Andrian had a chance to turn one of Gherman's favorite jokes upon him. After his mission in *Vostok 2,* Titov used to boast, jokingly, "I have lived seventeen days longer than any of you because I saw the sun rise and set seventeen times during my flight!"

Andrian, with scarcely a smile, said solemnly, "Now I have lived sixty-four days longer than you because I saw the sun rise and set sixty-four times during my flight!"

The girls were not satisfied to read the official reports of the flight and listen to dry lectures by the engineers and doctors on the mission. They wanted first-hand information. So they made a plan to kidnap Andrian and Pavel and hold them until they told everything.

As usual, after work, a number of the girls gathered in Valya's room, dressed informally in sweaters and shorts. Several of them lured the two cosmonauts into the room, and then Valya locked the door.

"Now, tell everything, right from the beginning, or you don't get out," she said.

Garrulous Pavel needed no urging.

"Well, I guess I will have to tell all," he said.

"Yes, you describe the flight to them, Pavlusha," Andrian said, moving out of the light and into a chair near the window.

"Well, as you know, my call sign in *Vostok 4* was Golden Eagle," began Pavel. "Just after I lifted off and began pulling a few g's, I had my first transmission from space. 'Golden Eagle, this is Hawk; how do you read me?' " he repeated, mimicking the scratchy noise of the radio. Hawk, of course, was Andrian, on his eighteenth orbit over the launch site.

"I forgot all about radio procedure," confessed Pavel. "I was so excited I yelled, 'Andrusha, Andrusha, I read you OK, and I am on the way up now.' "

As the two cosmonauts talked on, someone turned on the radio. Tchaikovsky's "Swan Lake" came softly on as background music for the narrative.

"I guess the high point of the flight, as far as I am concerned," said Pavel, "was when I passed within a mile or so of Andrusha."

"I'll go along with that," agreed the taciturn Andrian.

"I remember I did a funny thing," Pavel added. "I instinctively tried to dip my wings the way we do in a plane—but the *sharik* has no wings!"

They all had a laugh at this remark, including the usually somber Andrian.

"There was another thing I remember, too," said Pavel. "That was my first real weightlessness."

"I told you to go slowly and not be afraid," broke in Andrian.

"I should have paid more attention to you, Andrusha," Pavel said.

"I told you to unbuckle the straps slowly and to lift yourself gently out of the seat as we had been taught to do in the zero-g aircraft," Andrian went on.

"Well, girls, let me give you some advice; when you are in space and not buckled in, don't try to stand up as you would here on earth," Pavel said gravely. "It can't be done without a bang on the head, and I still have the bump to prove it."

The questions to the two were coming thick and fast. Pavel and Andrian, who had loosened up considerably, answered them almost without stopping to think.

"How did the earth look at night from space?" someone said.

"The most interesting sight, I think," said Andrian, "was the electrical storms going on over several parts of its surface."

"Yes," added Pavel, "it looked as if the clouds contained anvils and some giants were pounding on them, throwing off sparks that made the clouds flicker."

By now it was obvious that Andrian and Pavel had earned their freedom, and Valya agreed to release them since the ransom had been met.

In May 1963 the State Commission for Space Exploration met at the Academy of Sciences in Moscow. This important organization made all major decisions in the Soviet space program, including which cosmonauts would fly specific missions. Valery Bykovsky, Boris Volynov, Valya, and her alternate, a girl who is known only as Venera, were considered as primary and backup commanders for *Vostok 5* and *6*. Valery and Valya were tentatively, but not finally, selected.

The pace of Valya's training began to pick up. She no longer attended routine classes and exercises with her group. She and Venera began an extensive course in the various systems of the *sharik*. They went through the theory and operation of the electrical, communications, guidance, and propulsion systems. They spent hours inside the simulator that was an exact replica of the craft in which Andrian and Pavel had made their flights. Instructors on the outside presented them with problems suddenly to see if they took the right actions.

Andrian became Valya's self-appointed tutor for her work in the simulator. He would help her into the ejection seat of the *sharik*, and then very deliberately go through the steps of operating the Vostok.

"What is the first thing you do, Valya?" he would ask.

"First I run a functional check of all the systems," she would answer.

"Good," he would reply.

"I start on the port side of the *sharik*, with the pilot's control panel, then move on to the instrument panel; next I check the shutters on the VZOR device, and then I . . ."

"Yes, yes, that's enough of the checkout procedure," Andrian would cut in rather impatiently. "Show me how you perform manual control with the stick."

As he said this on one occasion he stuck his strong right arm through the open hatch and grasped the control stick with his short, blunt, tanned fingers. At the same time Valya also reached for it, and her small, white hand grasped his for an instant. Andrian turned deep red and yanked his hand from the cabin as if he had touched something on fire.

"Well, to pitch the *sharik* up, I pull back on the stick like this," Valya said, demonstrating.

"Good, that's right," stammered Andrian, "I am sure that you know how to manipulate it for roll and yaw control by now, so we won't go through those."

"If you say so," Valya grinned at the still-crimson Andrian.

Also included in the training were long sessions with Valery Bykovsky; his backup, Boris Volynov; and Venera. Valery and Valya, in particular, spent many long hours together going over and over details of the *Vostok 5* and *6* missions.

Despite the secrecy which surrounded Soviet manned space flight, news of events leaked out. Pavel, for example, goofed earlier in the year. On January 13, as he was boarding the plane in Havana to return to Moscow after a visit to Cuba, he said, "The world will soon know about the first female cosmonaut." Later in the year, on March 5, Radio Budapest announced that the USSR was preparing "new and sensational steps in space this year, including the first woman cosmonaut."

A break in the hectic pace of last-minute training came with the annual May Day holidays. Valya took a few days off to go back to Yaroslavl to see her family. Once there, she was under constant emotional stress. She knew that she would not see her family again before her space mission, yet she could not say a word about it. She had to maintain the "cover story" that she was training for the National Parachute Team. The inability to talk to her mother about her coming mission was all the more frustrating because Valya had always been extremely close to her.

Together they strolled along the bank of the Volga. Valya was more pensive than she had ever been in her life. Reflected in the river close to the bank was the blue May sky, dotted with white clouds and splashes of green from the trees. Farther out, toward the middle of the river, were mirrored pictures of belching smokestacks, racing trains, and tall brick buildings. Somehow the contrast reminded her of the two happiest periods in her life. The natural scenes brought back all the joy she had known as a girl in Maslennikovo. The industrial pictures reminded her of how she had grown to maturity and made her own way in life, first at the tire factory and then at Red Perekop.

Her mother knew that something was worrying Valya. She could feel empathetically that it had to do with her work in Moscow. However, she knew Valya well enough to realize that she would get nothing out of her until Valya was ready to talk.

"Valya, Valyusha, take care of yourself—and please eat more, you've lost so much weight," she said as she kissed Valya good-bye at the train station.

Remembering the old Russian saying that the traveler who turned his back would never return, Valya stood facing the station and her mother until the train was well out of sight.

On May 26, Yuri called Valery and Valya and their backups into his office.

"Well, this is it," he said simply. "Tomorrow you four are off for the pad, and—at least two of you—to space. But don't worry; there will be other flights, enough for all."

With the completion of her formal training, Valya was given the rank of junior lieutenant in the air force.

Following a custom established by Yuri himself, that afternoon Valery and Valya made a trip into Moscow. They silently walked through Red Square, past the gaudy Cathedral of St. Basil on to Lenin's Tomb. The guard admitted them at once, and the two went inside. Little did the other visitors realize that they were seeing two fellow citizens who in a few weeks would be circling far above Red Square in spaceships called *Vostok 5* and *Vostok 6*.

Sea Gull on the Steppes

Tyuratam, the Cape Kennedy of the Soviet Union, was opened in the summer of 1957, some twenty-five miles north of the small village of Tyuratam on the northern bank of the Syr Darya River. Located east of the Aral Sea in the semi-desert steppes of southern Kazakhstan, Tyuratam, which had only 8,000 people in 1957, is 280 feet above sea level, and its low, rolling terrain is dotted with wormwood and covered with reddish gravel. In January the winds howl out of the northeast, and tumbleweed gallops like Cossacks across the steppes. The temperature plunges as low as −33° F. Since the humidity in winter can be as high as 80 per cent, the weather is bitterly cold.

In May 1963, however, the temperature at Tyuratam was about 90° F., but the humidity was only 25 per cent. It was hot, but not unbearably so. In fact, conditions for launching *Vostok 5* and *6* were ideal.

On May 27, Valya, Valery, Boris, and Venera left Moscow for Tyuratam. With them were Sergei Pavlovich, Gagarin, Titov, and Nikolayev. Pavel Popovich did not accompany them because he was working in the mission-control center at Star Town near Moscow.

The trip was Valya's first long one on an airliner, and the Tu-124 plane was a good deal like the one in which she had practiced weightlessness. After leaving Moscow, the plane headed southeast toward the Aral Sea. Valya stayed by the circular window throughout most of the trip. She was fascinated by being able to see so much of her country. It was a patchwork quilt of colors. There were bright green rectangles surrounding countless collective farms, and large cities hidden beneath clouds of smoke. Here and there a large river ran like silvery Christmas tinsel dropped upon the earth, and the sun glinted off it as though someone were signaling with a mirror.

"How beautiful the earth is from the air!" she said, half to herself and half aloud.

"Wait until you have seen it from space, Valya," quietly said Andrian, who had been watching her since takeoff.

Occasionally, to rest her eyes, Valya would stop gazing out the window and rest her head on the pillow of her seat. At these times she would involuntarily glance at Venera. Usually her glance was returned. Both of them were thinking the same thought: *It is not really decided yet; the State Commission can still make a last-minute change.* But there was no envy between the two. Both were fully trained and equally ready for the mission. They had proven this beyond doubt throughout their training.

The Tu-124 touched down smoothly at the airport outside Tyuratam. The cosmonauts emerged from the plane into what at first felt like a blast from a furnace in a steel mill. The hot, dry air engulfed them. It was a shock after the pleasant weather they had left behind only a few hours earlier in Star Town. At the airport they were picked up in cars that drove them from the peasant village to Zvezdny Gorad, ("Star City"), which was still under construction in the desert to the north. On the way to their hotel, Valya noted the name of a movie playing in town. The title caused her to smile mysteriously; it was *A Concoction for Marriage.*

The hotel was reserved exclusively for cosmonauts, scientists, engineers, and those government leaders who were necessary—or at least on hand—for the launching. Valya and Venera checked into rooms on the first floor, while Valery and Boris went to rooms on the second floor with Yuri and Sergei Pavlovich. Also staying in the hotel were the newspaper, radio, and television reporters who were old hands at covering space-flight stories from Tyuratam. Sometimes they knew more about what was going on than anyone else, except, of course, Sergei Pavlovich.

The hotel had two unique features. One was the shrubbery around it. These pleasant spots of green in the drab desert had all been planted by cosmonauts who had spent a few days in the hotel before moving to the launch site for their last night on earth—for a few days! Yuri Gagarin had started the custom, and he had been followed in turn by Gherman, Andrian, and Pavel. Soon Valery and Valya would add to the growing number of plants. The other feature not found in usual Russian hotels was a private movie theater. It had a special purpose. The cosmonauts were restricted to the hotel for reasons of safety and security. Films, straight from Moscow, gave them one means of passing the time. There also was a special volleyball court where cosmonauts played reporters from *Pravda, Izvestia,* and Tass, or cosmonauts played cosmonauts, or cosmonauts and scientists and engineers played cosmonauts and reporters, etc. No matter who was playing whom, at least one man was never present. Sergei Pavlovich, if anyone could have found him on a rare day off, would not have been playing volleyball. But he could have been found at a secret spot on the Syr Darya River, where the fishing was really good. It was a place that even Andrian did not know!

Valery and Valya participated in the assembly and checkout of their Vostok spacecraft and booster rockets. Their presence in the assembly shops had a visible effect upon the workers and engineers preparing them. They also did last-minute training in the spacecraft itself, since there were no flight simulators at Tyuratam.

After a few days at Tyuratam, the State Commission met for its final decision, although the word had long ago leaked out that Valery and Valya would be the commanders of *Vostok 5* and *6*. However, the ceremony in the conference room of the Administration Building made it official.

Korolev made a short report, stating that in his opinion the spacecraft, boosters, and launching equipment were ready for the proposed flights. General Kamanin made a similar report, saying that the four proposed crew members were fully qualified for the flights.

"As commander of the *Vostok 5*," General Kamanin said, "we propose Valery Fedorovich Bykovsky."

He then added, "As commander of *Vostok 6*, we propose Valentina Vladimirovna Tereshkova."

Applause broke out through the room, led by the newspapermen who were present. The other cosmonauts, as well as the engineers and doctors, joined in. Valery and Valya, as was the Russian custom, applauded those who were applauding them.

Sergei Pavlovich stood up, and the noise immediately ceased. Even the newspaper reporters became quiet.

"Our nation," he said, "can truly be called the shore of the universe. I believe that from this brilliant shore new spaceships will leave for outer space. I congratulate Valentina Vladimirovna on her appointment as commander of the spacecraft. I say this with genuine excitment. Valentina Vladimirovna is the first woman in the world who will pilot a craft in outer space."

On June 14, just two days before Valya was scheduled to lift off, an Italian newspaper carried a story that both surprised and infuriated General Kamanin and the party apparatus. Obviously, someone had talked. The *Corriere della Sera* reported that a Soviet woman cosmonaut had been launched in a dual flight. The story was based upon supposed radio intercepts by two brothers named Giudica-Cordiglia in Turin. Other papers quickly copied the story and elaborated it. One named the woman cosmonaut as Ludmila and said she was twenty-five years old. The age was close, but the real name was Valentina!

On the evening of the day before Valery's launch, Sergei Pavlovich invited Valery and Valya down to the launching pad to watch the countdown on *Vostok 5*. He was dressed, as usual, informally, in a bright sports shirt not tucked into his gray trousers. A floppy straw hat kept the fierce desert sun off his thinning gray hair. The warm wind blowing off the steppes brought a faint hint of wormwood and an even fainter aroma of thyme from the bushes along the pad and fluttered the red warning pennants that decorated the huge rocket. Its gleaming white propellant tanks could be seen here and there through the trusses, girders, and beams of the launch umbilical tower, support arms, and gantry. A large corrugated metal pipe snaked its way up to the *Vostok 5* from the air-conditioning unit on the ground.

Wisps of escaping oxygen from the rocket drifted among the girders and lines.

As usual when the Chief Designer was present, engi neers and technicians descended upon him waving drawings and reports and asking questions. He calmed them down and assured himself that everything was going smoothly. He then told Valery and Valya that he had a little time in which to talk to the newspaper reporters and that he felt he had to.

"Well, let's get on with it," he said impatiently.

"This dual flight of *Vostok 5* and *Vostok 6* has been planned for a long time. We need the information they will provide for orbiting space stations of the future as well as the spacecraft that will take men to the other planets.

"Until now space journeys were made by jet pilots, a tough breed accustomed to high speeds and high accelera-tion loads. Valentina Tereshkova, commander of the *Vostok 6*, has become familiar with these loads during her training program. The flight of Valentina Tereshkova and Valery Bykovsky is another step forward, both as regards the dura-tion of the flight and the research to be carried out—re-search such as astronomical and geophysical observations."

"In speaking of extending the boundaries of space flight," broke in one of the reporters, "do you have in mind flights to the moon?"

"I certainly do," Korolev answered immediately, "but it must be pointed out that a moon flight, though a very fas-cinating undertaking, is also a very difficult one. But I am sure that the time is not distant when man's first flight to the moon will become a reality, though more than one year will be needed for the solution of this problem."

With these final words, he turned abruptly from the reporters and went back to his office. His sixteen-hour day still had several hours to run.

Korolev felt that the winds aloft were too great and he postponed the launch of *Vostok 5* by one day. The cosmonauts spent it on the traditional fishing trip and picnic that had been started by Yuri Gagarin. They and their friends piled into a truck and drove a considerable way through the steppe before stopping at a secret fishing place on the Syr Darya River. Since Valery and Valya were going into space, they were given the menial jobs. They were cook's helpers and cleaners-up. Yuri appointed himself "director of ground operations," and as such, chef for the day. He made Andrian and Gherman responsible for catching the fish, which they did expertly. Just as expertly, Yuri cooked a special soup, for which Valya and Venera peeled the potatoes.

One of the games played after the picnic was King of the Mountain. In this game, the king was tossed into the river with his clothes on. He turned out to be a very high ranking scientist with one of the most respected titles in the Soviet Union: doctor of technical sciences. One of the cosmonauts noticed this as he handed the nearly drowned young man his waterlogged wallet.

Valya and the other cosmonauts, with their still-shivering doctor of technical sciences, did not get back to the hotel in Star City until just before dawn. The picnic, obviously, had been a success.

June 13, the day before Valery was to lift off, was a hectic one for all. They spent a last session together going over the radio procedures that would be used in the *Vostok 5* and *Vostok 6* missions, practicing both voice and code communication between the spacecraft and the launch-control center. In the afternoon Valery made the traditional trip to the launch pad to meet the launching crew and accept his booster and *Vostok 5* spacecraft. The members of the crew from the launch-control center presented him with a bouquet of flowers (not easy to come by on the steppes of

Kazakhstan). An air force band played the national anthem of the USSR, and a round of speeches by various engineers and dignitaries from Moscow began.

Valery was clearly embarrassed, but he shook hands and chatted with the technicians who would be responsible for launching him into space on the following day, Then Valery accompanied by Sergei Pavlovich, General Kamanin, Valya, Boris, and Venera, and a retinue of others, made an inspection tour of the launcher and the rocket. Both Valery and Sergei Pavlovich were pleased with the way the countdown was going and the condition of the rocket and spacecraft.

That evening, Valya and Venera accompanied Valery and Boris to the cosmonauts' cottage. On the night before a launch, the cosmonaut and his backup leave the hotel in Star City and spend the night at the launch site in a quaint little wooden cottage with a slate roof. It is set among a few poplar trees and surrounded by acacia shrubs. The cabin is only large enough for the two cosmonauts and two doctors. There is one bedroom with two beds for the cosmonauts, another bedroom for the doctors, and a bathroom. The living room is simply furnished with a table, radio, television set, and magazines and newspapers. Also a small table with a chess set provides for a favorite recreation.

This was the second night that Valery was to spend in the little cottage. He had spent the night of August 10, 1962, there as backup cosmonaut to Andrian during the *Vostok 3* mission. This time, after he had left the cottage in the morning, his portrait, by custom, would be placed on the wall over the bed in which he had slept, added to those of Gagarin, Titov, Nikolayev, and Popovich. Two days later, the first woman's picture would appear there too.

Inevitably, the newspaper reporters made their way to the cottage and demanded to talk with Valery. At this point,

Valya and Venera said good night to Valery and returned to their hotel rooms in Star City. They would be back the following night, however.

Launch day, June 14, was clear and sunny. Only a few wisps of clouds were scattered across the wide blue sky. Valery and Boris did not show up for the morning exercises with Valya and Venera and the other cosmonauts. But Valya heard that they had both spent a restful night and were now getting into their space suits. She and the others would accompany them to the launch pad in the blue bus, especially outfitted for carrying cosmonauts there from the medical building.

The bus drove right to the foot of the launcher, so Valery had only a few feet to move in his bulky, bright orange space suit. He stopped briefly at the foot of the steps leading up to the elevator and shook hands with Sergei Pavlovich. He said good-bye to Andrian, Boris, and Yuri. Then he turned to Valya: "Until we meet in orbit." Valery went up the steps and entered the elevator. It slowly took him up to the work platform surrounding the *sharik,* and he leaned over the railing to wave to Valya and the others on the pad below.

After the launching pad had been cleared, Valya and the others got back in the bus and drove over to the two-story observation post some three miles away. Through the loudspeaker they heard Valery talking to Yuri, who was *sharik* communicator for the launch-control center. Valery was calmly reading off the conditions of the cabin temperature, air pressure, humidity, etc. Yuri was checking them against the information being received by telemetry. Everything was proceeding just as it should. Other familiar voices came over the loudspeaker.

"This is Falcon, Hawk," said Andrian from the blockhouse beneath the pad. "How do you read me?"

"I read you fine, Falcon," replied Valery.

"This is Golden Eagle, Hawk," said Popovich from the mission-control center in Moscow. "How do you read me?"

"Loud and clear, Golden Eagle," said Valery.

Suddenly the voice of Yuri Gagarin broke in, "We are now at T minus fifteen minutes and counting."

Valya took a seat behind a pair of powerful binoculars mounted on the railing around the observation stand. Looking through them, she suddenly seemed to be on the pad itself. Through the shimmering heat waves, she barely saw the rocket and *Vostok 5* enclosed in the maze of the launcher's girders and beams. Here and there she saw faint wisps of oxygen drifting from the launcher as the propellant tanks of the booster were being "topped off."

"T minus five minutes," said Yuri over the intercom speaker behind her.

At this point, the two halves of the gantry tower folded back onto the pad, and then the umbilical tower supplying electric power and propellants to the rocket pivoted back. The rocket was left gleaming white in the hot sunlight, held erect by four support arms. It was now on internal power, ready for launch.

"Falcon, this is Hawk," came Valery's voice over the intercom. "Ready for launch; helmet closed; gloves on and secured!"

Valya moved back from the binoculars to let Venera have a look. Boris had joined them after taking off his space suit and changing into sports clothes. He also had a look at the huge rocket that almost had been his.

"Ignition, we have ignition," said Yuri.

The flame deflector beneath the launcher shot a tornado of white and orange flame and smoke out across the desert. Dust and smoke boiled up from the pad itself, obscuring the base of the rocket with its twenty-four main engines all belching flame at the same time. In less than two seconds, the four support arms pivoted back, releasing the rocket. No one heard Yuri's voice over the intercom saying, "Lift-off! We have lift-off!"

Slowly, lazily, the rocket inched upward from the pad. As it grew lighter, it picked up speed. With smaller binoculars Valya, Venera, and Boris followed it until it was some thirty miles up. There they saw the four strap-on stages separate and tumble back to the desert.

After Valery was inserted into orbit, Valya left the observation post with Venera and Boris and went into the control center to meet Andrian. The place was in its usual uproar during the early phases of a mission. Phones rang incessantly. Technicians and engineers bent over tables containing charts and long columns of figures from the computers. Sergei Pavlovich, as always, was the focus of all this activity. He was never alone for a minute. There was always a decision to be made.

Valery's face was on the television screens of several of the monitors, and Yuri and Andrian kept in constant touch with him while he was over areas of the Soviet Union where there were tracking radars.

Feeling absolutely useless, Valya watched the proceedings until Valery's twentieth orbit. Then she and Venera left to attend her acceptance of her rocket and spacecraft.

Valya selected a simple blue dress. With it she wore white high-heeled shoes. Venera picked out a coral dress that went with her black hair. The familiar ceremonies took place as the pad itself was still warm from the launching of *Vostok 5*.

"I really envy you, Valya," sighed Sergei Pavlovich. "Just imagine, being able to go into space. I am always left behind on the pad."

"Don't worry, Sergei Pavlovich," Valya said with a

grin. "When I get back from my trip, you and I will plan one together—just the two of us. Where shall we go? To the moon or maybe Mars?"

The band played the national anthem, people from everywhere gave her bouquets of flowers, and she made a little speech thanking them all.

Then she did something unusual for the protocol of the ceremony. She turned and presented her largest bouquet to Sergei Pavlovich. "These are for you from the bottom of my heart," she said.

For one of the few times in his life, Korolev was almost completely speechless. "Thank you, Valentina Vladimirovna," he replied very formally.

With the ceremony over, Valya and Venera changed into slacks and returned to the pad with Sergei Pavlovich. The three of them took the elevator up to the work platform around *Vostok 6*. There they were met by a very intense young man with curly hair, wearing wire-frame glasses. He was the spacecraft engineer. He knew every system in the *sharik* perfectly, in greater detail even than the Chief Designer and the cosmonauts. It was he who saw to the fastening of the hatch after the cosmonauts had entered the *sharik,* and it was his responsibility to ensure that the seal was a good one.

First Valya and then Venera climbed into the ejection seat. Sergei Pavlovich had each of them go through the launch procedures until he was satisfied with the answers.

On the way from the pad, he relaxed considerably. He no longer talked about the Vostok or its booster.

"Valya, when was the last time you wrote to Yelena Fedorovna?" he asked suddenly, in a tone like the father that Valya had never really known.

"It's been too long, I am afraid," she replied, feeling very guilty.

"Then be sure to do so as soon as possible," he said assuming his familiar, gruff mumble.

And this she did, being careful not to say where she was or what she was doing.

The remainder of the afternoon was spent with the omnipresent and persistent doctors and medical technicians.

After the doctors had finished drawing blood, listening to hearts, poking and prodding, and asking interminable questions, Valya and Venera left the medical building and returned briefly to the launch-control center, accompanied by one of the two doctors who would spend the night with them in the cosmonaut cottage.

Things were going well with Valery. He had gone through several important experiments, including free-floating in zero gravity.

"Falcon, this is Hawk," Valery's voice crackled over the intercommunication speaker in the room. "Is Sea Gull there?"

"Hawk, this is Falcon," the capsule communicator replied. "She just walked in."

"This is Hawk," Valery said. "Pleasant dreams, Sea Gull. I'll be waiting for you tomorrow, in space, Valya."

At the cottage, they found the housekeeper waiting for them. "Valentina Vladimirovna, you will occupy the bed there—the one that Gagarin and Nikolayev used," she said rather formally.

Above the bed, Valya noted the portraits of those two predecessors. She glanced at the other one, in which Venera would sleep. Hanging on the wall above it were portraits of Titov, Popovich, and Bykovsky. In the morning, after she left for the medical building, Valya's portrait would hang with those of Yuri and Andrian.

After tea, Sergei Pavlovich and Yuri Gagarin dropped by the cottage. There were no newspaper reporters this time. Sergei Pavlovich had seen to that!

This visit was a new experience for Sergei Pavlovich. He was used to a quiet walk and "shop talk" along the gravel path among the birches with men like Gagarin, Titov, Nikolayev, and Popovich. He wondered what he could talk about with two pretty girls.

Valya solved his problem for him. She served him and Yuri tea, and then they sat on the tiny porch of the cottage. Valya sang an idiotic song that the cosmonauts and the newspaper reporters had made up in the weeks preceding the mission. Some of the verses had appeared in the cosmonauts' newspaper *Neptune*.

Sergei Pavlovich did not know what to think about this behavior. The tune was familiar. It was "Our Little Town Isn't Much." However, the words were all wrong and made no sense to him. One verse ran:

The reporters are digging in the bushes;
The reporters are searching here and there.
Where will Gagarin lead
The woman cosmonaut in the morning?

He shrugged, and turned. "Yuri Alekseyevich, let's leave these two beauties; one of them has a full day tomorrow," he said.

Yuri wished them luck, as did Sergei Pavlovich, and they left—Yuri for the hotel in Star City and Sergei Pavlovich for his office; his sixteen-hour day still had quite a few more hours left. One of the things he needed to check was the data received from the sounding rockets that were being launched periodically to monitor radiation in the regions through which Vostok would orbit.

While Valya and Venera slept, the housekeeper entered their bedroom. On Valya's night table she placed a bouquet of white gladiolas and daisies, as she had done during the *Vostok, 1, 2, 3,* and *4* missions. She also noted that the electric fans on the table near the open window were blowing directly onto Valya and Venera. Quietly she switched them off.

Neither Valya nor Venera knew that the doctors in the next room made an hourly check to see how they were sleeping. They were also unaware that on two occasions during the night Sergei Pavlovich looked in on them and received a report from the doctors.

June 16 at Tyuratam was a perfect day. There were few clouds in the sky and no ground winds that could cancel the launch. But it was hot. The temperature would reach the mid-nineties by lift-off.

Valya and Venera were awakened early in the morning by one of the doctors.

Both drifted back to sleep, especially Valya. She had never been known to spring out of bed when the sun first peeked over the horizon.

"Girls! Girls!" the housekeeper called. "You've got to get up; it's time!"

For a moment Valya thought it was Yelena Fedorovna calling to her and Lyuda. For a moment she expected to hear the familiar "Children, get ready for school! The rooster has crowed long ago. . . ."

After a few setting-up exercises and a shower (Valya decided not to put on a dab or two of her favorite "Red Moscow" perfume) , the young women sat down to a breakfast of space food. It consisted of veal cutlets, bite-size caviar sandwiches, wheat bread, lemon slices, and coffee with milk. For dessert there was a piece of candy that was really a vitamin pill. The meat, sandwiches, bread, and lemon slices were all sealed in plastic envelopes. The coffee was in an aluminum tube like toothpaste.

The preparation for space was a long and tedious one. It began with the attachment of the biomedical sensors to the young women's bodies. These small disks would detect changes in the electrical activity of various organs or muscles. The changes would then be amplified and transmitted to earth by the Vostok's telemetry system. Valya had three sensors to measure the electrical activity of her heart. Another sensor, to measure the mechanical activity of her heart, was also attached to her breastbone. Four sensors were placed on her forehead and scalp to measure brainwave activity. Two more were stuck to her right shin and the sole of her right foot to measure changes in the skin's electrical resistance. Four more were placed at the corners of her eyes to record eye movment, so that doctors could gauge the depth of her sleep.

Looking more like telephone switchboards than cosmonauts, Valya and Venera then slipped into their specially designed undergarments—seamless, knitted sweaters, long drawers, and seamless paper socks. The final sensor was attached after they had put on the underwear. It was an elastic band around the chest that measured the rate of breathing.

Next came the blue pressure suit itself, which actually kept the air against Valya's body. It was a struggle to get it adjusted properly so that there were no pressure points on the body and no binding. Valya's suit was different from that worn by Gagarin, Titov, Nikolayev, Popovich, and Bykovsky. It had a special emblem sewn on the left shoulder: a white dove on golden sun rays, with the red letters "USSR." Finally, over the pressure suit, came the bright orange flight suit, which was worn so that Valya could easily be spotted in case she came down in the water. Dressed at last, Valya did something the previous cosmonauts may not have done. She checked herself in a mirror to see how she looked!

As did each of the cosmonauts who had preceded her

into space, Valya took along some personal mementoes of her flight. These were small replicas of the flag of the USSR, which she would distribute at the World Women's Congress in Moscow that would take place after her mission.

Sergei Pavlovich checked in with the two cosmonauts just as they completed suiting up. He told them that everything was ready on the pad and that the countdown was still running.

Valya and Venera struggled into the blue bus for the trip to the pad, helped by Gherman and Andrian. Once in the bus, pandemonium broke loose. Everyone started talking at once. As usual, it was crowded with cosmonauts, doctors, and other technicians. In fact, there was so little room that Gherman and Andrian had to sit on the floor near Valya's feet. The chattering and singing kept up all the way to the pad.

When the bus came to a stop near the launcher, Valya had a sudden compulsion to lean over and kiss Andrian's dark, wavy hair. However, she had to settle for a very formal handshake instead.

"Say hello to Valery for me," he said.

Valya leaned over and clinked her space helmet against that of Venera. It was the traditional farewell from the prime cosmonaut to the backup.

She moved clumsily in the bulky suit, carrying in one hand the portable air-conditioning unit which kept her suit at a comfortable 68° F. while it was 95° F. on the pad. She shook hands gravely with Konstantin N. Rudnev, chairman of the State Commission for Space Exploration; with General Kamanin; and with Mstislav V. Keldysh, president of the Academy of Sciences of the USSR and chief theoretician of cosmonautics. Sergei Pavlovich gave her a bear hug and wished her good luck. Valya made her formal speech and then moved awkwardly up the stairway to the elevator.

Just before blast-off, June 16, 1963.

Titov and the spacecraft engineer accompanied her to the work platform surrounding the Vostok. Before entering the *sharik,* she paused a moment and clasped her hands above her head like a boxer, in farewell to those on the pad 104 feet below. Then she took a deep breath of the warm air sweeping over the steppe, fragrant with faint hints of wormwood and thyme.

Gherman and the spacecraft engineer lifted her up and removed the plastic overshoes from her feet. Then they gently lowered her into the ejection seat. For a few moments they chatted with her, and Valya became absorbed in her checkout procedure. She did not even know that the hatch with its explosive bolts was being sealed and inspected by the spacecraft engineer.

"T minus fifteen minutes, Valya." Yuri's voice came through her headphones.

She continued the checkout automatically, as she had done so many times before in the *sharik* simulator.

"T minus ten minutes," Yuri said.

"Valya, I wish you a happy flight," Sergei Pavlovich's voice broke in.

"T minus five minutes," Yuri continued.

"Right, I have closed my helmet and put on my gloves," Valya reported. Engineers in the launch-control center read their meters. The suit was pressurized and holding steady at 5.7 pounds; the cabin temperature was 68° F. The carbon dioxide was only 0.15 per cent. The cabin pressure was 14.5 pounds.

At 12:30 A.M., June 16, 1963, Valya heard Yuri's voice in her earphones above the muffled roar of the rocket engines: "Lift-off! We have lift-off!"

"It Is I, Sea Gull!"

The huge booster suddenly shuddered into life. Even inside the Vostok under its aluminum nose cone, the noise reached the same level as that made by air hammers tearing up a street in Moscow. However, Valya wore special earphones and protectors that reduced the level considerably. With this noise also came a vibration that made it impossible to do anything but sit strapped in the ejection seat and jiggle uncomfortably in all directions at once.

Then the acceleration began, just as it had when the devil's merry-go-round had begun to move.

As the rocket burned off its liquid oxygen and kerosene, it shot upward faster and faster. An invisible giant hand pushed Valya into the seat. Her breathing became very difficult, and her heart rate climbed to 154 beats per minute, which was only two beats faster than Valery's had been at lift-off. After two minutes, the four strap-on boosters shut down and dropped away. The acceleration let up somewhat, and Valya's heartbeat dropped to 148 beats per minute. Her breathing became easier, too. Suddenly, there was a ping, and the explosive bolts blew the nose cone off.

Finally, the core vehicle shut down and the third stage ignited to send the Vostok into orbit. The acceleration produced by the third stage was nowhere near as great. Valya felt as though she were riding in a car that had been hit from behind by another moving car. Even so, her pulse was still 140. During this period the doctors in mission-control center anxiously watched her pulse rate being traced in squiggly inked lines on the strip recorder that was converting the signals sent from the telemetry system to a graph that they could read. They noted that the rate was high, but they decided that there was no need for alarm.

Gagarin's rate had reached 149, but imperturbable Andrian's pulse had been only 130. Both were higher, however, than those of the first American astronauts. John Glenn's pulse had only been 110, less than 40 above normal, and lighthearted Wally Schirra's had only reached 112.

Through her right porthole, Valya could see the earth.

"It is I, Sea Gull," she excitedly called to mission-control center, forgetting her carefully instilled radio procedure. "Everything is fine . . . I see the horizon; it's a sky blue with a dark strip. How beautiful the earth is! Everything is going great," she chatted away.

"Sea Gull, this is Dawn," said Yuri, correctly following procedure. "Everything *is* fine, Valya; *Vostok 6* is right on the orbit planned."

"Hawk, this is Sea Gull," she radioed. "Do you hear me?"

"This is Hawk," Valery replied. "I hear you clearly, Sea Gull. Welcome to space, Valya!"

Vostok 6 was at apogee, or farthest from earth, at 143 miles and was at perigee, or closest to earth, at 112 miles. She was traveling at 18,000 miles per hour in an orbit that crossed the equator at an angle of 65 degrees. She completed an orbit every 88.3 minutes. However, Valery's spacecraft was in a different orbit. His apogee was 129 miles, and his perigee was 104 miles. It took him 88.06 minutes to make an orbit, and his orbit was also inclined at an angle of slightly more than 65 degrees to the equator. Since Valery's spacecraft had separated prematurely from the booster, it went into an orbit with a lower perigee than was planned. There was not enough time to reprogram the guidance unit of the ground computers to place the *Vostok 6* in an orbit that more closely matched Valery's. Thus, the two spacecraft could not rendezvous. However, the differences in their orbital angles meant that they would pass within about three miles of each other.

After a quick check of the instruments in the *sharik*, Valya noted that everything was well within limits. The temperature was a pleasant 75° F. and the relative humidity was only 37 percent. The air pressure was one atmosphere, or the same as at sea level on earth.

One of her first tasks was to make an entry in her logbook of her initial impressions and sensations.

"I withstood the powered phase of flight very well," Valya wrote. "I was a bit nervous. After the separation of the booster, the transition to weightlessness was very smooth and there were no problems, perhaps because I was so busy at the time. I did not notice any problem with my vestibular apparatus."

She wrote on, "In weightlessness it is possible to work, and I am in good humor, especially after talking to Hawk on the radio."

The problem that caused most concern to the space doctors and biologists was one that had first manifested itself on Gherman Titov's *Vostok 2* mission. During his sixth and seventh orbits he became extremely nauseated and dizzy. The doctors attributed it to sudden head movements while he was weightless. The movements under this condition affected his vestibular apparatus—the semicircular canals and otolith organs of the inner ear, which together supply information to the brain that provides a sense of balance.

Valya was assigned several figures to draw at specified times during the flight. One of these was a spiral; another was a five-pointed star. She also had to make parallel lines at right angles to each other. A further test consisted of the familiar trick of closing the eyes and then touching the tip of the nose with an extended index finger. In Valya's case she had to perform it while a television camera stared at her, and doctors clustered around their monitoring sets in the mission-control center at Star Town. Signals sent from the sensors taped to the corners of her eyes also indicated whether or not they were rolling about in a condition called nystagmus, which occurs with dizziness and disorientation.

Valya passed the tests with no problems—and, indeed, never became dizzy or nauseated during her mission. Neither did Valery, an experienced jet pilot.

Gradually the excitement subsided. The doctors in the mission-control center noted from the paper charts flowing from the strip recorders that Valya's breathing rate had dropped to 20 and her pulse rate to 80, both of which were near her normal rates on earth.

She continued to follow the mission plan, monitoring the condition of her ship and transmitting information to earth. In less than an hour, she plunged from day into night. She looked first out of the porthole on the right, from which she could see the earth, and then out of one on the left, through which she saw the sky. Through the left porthole she saw such familiar constellations as the Big Dipper. There was a difference, however. The individual stars burned like clear electric lights. There was no flickering and twinkling, as there was when she saw them from the earth. Through the right porthole she saw the surface of the earth rolling rapidly beneath her. It was dimly lit by the moon, but she could see patches of clouds clearly.

Suddenly a burst of sunlight shot through the left porthole and the number 1 appeared on the orbit-counter meter of her control panel. Night was over, and so was her first orbit in space.

She made a special report on her mission to Premier Nikita Khrushchev. In return, mission control read her a special telegram that the premier had sent to her:

> DEAR VALENTINA VLADIMIROVNA, MY HEARTIEST CONGRATULATIONS TO YOU, THE FIRST WOMAN COSMONAUT IN THE WORLD, FOR A REMARKABLE FLIGHT IN OUTER SPACE. THE SOVIET PEOPLE ARE PROUD OF YOUR TRIUMPH. WE ALL FOLLOW YOUR HEROIC FLIGHT WITH GREAT ATTENTION AND FROM THE BOTTOM OF OUR HEARTS WISH YOU GOOD HEALTH AND GOOD LUCK AND A SUCCESSFUL COMPLETION OF THE FLIGHT. WE WILL MEET YOU WITH GREAT PLEASURE ON OUR SOVIET HOMELAND.

One of the most beautiful sights she saw from the right porthole was the halo around earth that would capture the attention of cosmonauts and astronauts for over a decade to come. Yuri Gagarin had told her how beautiful it was, but she simply could not imagine it. Now she saw it. From the earth outward, it began with a light blue that delicately changed into turquoise, then into dark blue and then violet, and finally disappeared into the blackness of space.

Her first thought on seeing it was that she wished she could find a material with such colors for a dress.

Another night view that never failed to impress her occurred each time she orbited over America. The large cities with their rays of streetlights looked like diamonds on the earth. One city—she did not know which one—had a circular freeway around it that she could see clearly. The string of light around it reminded her of a pearl necklace. On one night, she was observed orbiting over Tyuratam. The capsule communicator on duty called, "Sea Gull, Sea Gull! We can see you overhead."

"This is Sea Gull. I hear you well, and I see the lights of the center. I see your lights!" said Valya excitedly.

The earth in daylight was equally beautiful. Each continent had its own color, she noted. Africa was yellow, except for the blue ribbon that was the Nile. South America was green. Asia was a dark brown, except for the green arrowhead jutting into the Indian Ocean. The waters of the Atlantic Ocean were darker blue than those of the Pacific. The American Rocky Mountains were dark gray, but the Himalayas were reddish and covered with white spots.

As luck would have it, when she orbited over Yaroslavl, the earth below was covered with clouds. The largest of the Indonesian islands looked like emeralds against the light blue of the Indian Ocean. There was no mistaking Greenland's stark whiteness in the dark blue of the North Atlantic Ocean and Arctic Ocean. Between earth and sky below her she could see the condensation trail of high-flying airplanes.

At each sunrise, Valya, as well as Valery, reported seeing "a mass of small particles through the right porthole." They looked, she said, as if the spacecraft were passing through a meteoroid cloud. Valery was a little more precise. He said that they were small illuminated dots at a distance of two to six feet outside the right porthole. These were, of course, first reported by American astronaut John Glenn and later seen by Scott Carpenter. Later they were determined to be ice crystals from the life-support system of the spacecraft.

Before she knew it, during the third orbit, the spacecraft's log called for her to eat lunch. Valya had an idea: Why should she not dine with Valery?

"Hawk, this is Sea Gull," she radioed. "How about joining me for lunch?"

"Sea Gull, this is Hawk," came the reply. "Sorry, Valya, but I have about two hours to go."

Suddenly Valya remembered. She had been launched three and a half hours earlier in the day than Valery had. Thus there was a difference of that many hours in their work and dining schedules.

"Sea Gull, this is Hawk," Valery called. "Have a good lunch, anyway. By the way, what's on the menu?"

"Hawk, this is Sea Gull," she replied. "I am having herring and egg patties, white bread, fresh apple, and black currant juice."

"Sea Gull, this is Hawk," Valery returned. "It sounds good, but I am having roast veal for the main course."

This meal was typical of those that she would eat during the course of her mission. Other main dishes included roast beef, roast tongue, chicken fillet, red caviar, rice and eggs, and curds. In addition, there were fresh orange, lemon, and apple slices. For drinks there was a choice of cherry juice, black currant juice, coffee, or tea. During the mission Valya consumed about 2,529 calories per day while eating four meals.

She also had one quart of water a day for drinking purposes. It was especially "blended" for her taste and contained chemicals to keep it fresh. She drank it through a plastic tube connected to the water tank.

One convenience the *sharik* did not have that Valya missed before the end of her mission was a shower. Instead she had to rely on specially treated washcloths. These were moistened with glycerine and a mild detergent. At least she did not have Valery's problem: the discomfort of having to go without shaving.

During the fourth orbit, Valya received a surprise call on the radio.

"Sea Gull, this is Dawn," said Yuri. "We have someone who wants to talk to you."

"I hear you well. Your call sign is Sea Gull, I believe," said Nikita Sergeyevich Khrushchev. "Permit me to call you Valya, Valentina. I am very happy, and it gives me fatherly pride that our girl, a Soviet girl, is the very first in the world to be in outer space and to be the master of a very advanced technology."

"Nikita Sergeyevich, I am very excited and touched by your message. Many thanks for your concern and your kind words," replied Valya.

They chatted for a few more minutes.

"Again, many thanks for your thoughtfulness and kind words, Nikita Sergeyevich," Valya concluded. "Until we meet again on earth, good-bye."

During the sixth orbit, Valery called Valya's attention to an interesting phenomenon on the darkened earth below. It looked as if someone were signaling to them from the Indian Ocean. Clouds winked on and off as if there were light bulbs inside them or as if they were shapeless neon signs.

"It's a thunderstorm," said Valery. "But don't worry; there's no rain or lightning up here."

This sight made such a great impression on Valya that she wrote about it in her logbook. "On the sixth orbit I saw a storm over the Indian Ocean. The sky was lighted by bright flashes. The nocturnal horizon was rather uniform, even before sunrise. Before each sunrise there is a unique sight. The clouds over the ocean have the form of ridges, and more often, of streets with small breaks in them."

Despite the excitement of the day and the mission so far, Valya had begun to get a little drowsy. She began nodding. Finally, she drifted off into an unscheduled and unauthorized nap. She failed to hear the appeals to Sea Gull from Dawn coming through her headset.

"Hawk, Hawk, this is Dawn. Wake up Sea Gull; she must be asleep," went out the call from Sergei Pavlovich to Valery.

Valery finally managed to wake her up.

Her first day in space, and Valya had gone to sleep on the job!

During the seventh orbit, the schedule called for Valya to go to sleep. Although she already been up for six "nights," it was only twelve hours since she had awakened in the cosmonaut's cottage at Star City! Her eyes ran over the meters on the instrument panel—*sharik* temperature 68° F., pressure right at one atmosphere, and humidity 34 per cent. Everything seemed right for a good night's sleep. She adjusted the temperature control to lower the cabin to 64° F.

Valya turned off the lights of the *sharik* and drew the blinds over the two portholes. There was only a faint glow from the phosphorescent numbers and hands of the meters on the instrument panel. The noise level in the *sharik* had dropped to 76 decibels, or about that found in conversation with someone at a distance of three feet. Most of the noise came from the droning of an electric fan which circulated the air through the cabin. Unfortunately she could not cut it off. Since even the air in an orbiting spacecraft is weightless, there is no natural convection to keep it in motion. The fan was used to keep it constantly stirred up so that it could be reprocessed by the life-support system of the spacecraft.

After calling Hawk and wishing him a good night, Valya adjusted herself in the ejection seat. Then she tucked her hands underneath the shoulder belt. Gherman Titov had told her that if she did not do this, her hands and arms would float weightless above her and interfere with her sleep. Thus settled down for the night, she closed her eyes.

It had been a long day.

During it she had traveled 180,000 miles since leaving the launching pad at Tyuratam. It had also been a long time since she had seen Andrian. Suddenly she missed him very much. Once the mission was over, they were going to be married. She wondered what Venera and Yuri and the others would say if they knew—or Mother, for that matter. They had not told a single person that they were engaged.

With Andrian on her mind, she quietly went to sleep.

The doctors in the mission-control center crowded around their strip recorders.

From the traces on the EEG (electroencephalogram), the doctors could tell that Valya was asleep.

As they watched their instruments throughout the night, they noted from the signals being traced out on the paper by the EOG (electrooculargram) that she did not dream. They also noted, with satisfaction, that her pulse dropped to 52. Her breathing rate was a steady 21, which was also good.

Valya awoke at 6:10 A.M., after sleeping through the twelfth orbit. She pulled her arms from beneath the belts and stretched, as far as she could while strapped into the seat. The first thing on the schedule was a fifteen-minute period of setting-up exercises. Despite the fact that she was weightless, she performed a series of specially devised isometric exercises which pit one muscle against another. Thus she was able to keep her muscle tone during weightlessness, although such exercise did nothing to keep her skeletal system in shape.

Following the exercise period, a "bath." Valya ran the moistened washcloth over her face, taking care not to loosen the EOG sensors taped to the corners of her eyes and the EEG sensor on her forehead. She washed her hands and went as far up the wrists as she could manage with her gloves off. Then she ate breakfast.

Today was going to be a very important one for Valya.

Before she had been launched, Sergei Pavlovich told her there was a possibility that her mission would be extended to three days instead of the planned one day. He said that it all depended on her condition and the condition of the *Vostok 6*. The State Commission for Space Exploration would make the final decision sometime Sunday night.

"Sea Gull, this is Dawn," came a familiar voice. "How do you feel this morning?"

"I feel . . . are we 'go' for two more days?" Valya asked bluntly.

"Yes, we are agreed," said Sergei Pavlovich. "You are 'go' for two more days. That will mean a total of forty-eight orbits rather than thirteen."

Hawk reported that he had heard the news and was happy to know that Sea Gull was going to stay in space and land with him.

On June 17, Valya performed a rather simple experiment that was to have far-reaching scientific effects. As *Vostok 6* orbited through earth's umbra, or shadow, over the Atlantic Ocean, mission control told her to take the Konvas movie camera from the stowage compartment and be prepared to aim it through the right porthole.

"Sea Gull, this is Dawn," came the scratchy voice through her headset. "Valya, we want you to aim the Konvas through the window now. Place it right on the horizon when it appears and start shooting."

"Dawn, this is Sea Gull," Valya replied, very businesslike. "I understand."

What Valya had done was to photograph two faint dust layers in the stratosphere, some eleven miles above the earth's surface. Measurements had been made from below, using balloons and airplanes, but Valya made the first pictures from above the atmosphere looking down into it. From her pictures, scientists were able to compare data from earth with data from space and compute the density of these two layers.

As they continued to orbit earth, Valya and Valery sent messages they jointly composed to the various countries as they passed over them. In turn, Valya received a message relayed to her from the mission-control center.

> MY DEAREST DAUGHTER VALYA, I AM VERY HAPPY THAT I HAVE SUCH A BRAVE AND COURAGEOUS GIRL. I AM VERY PROUD OF YOU, AND I CAN HARDLY WAIT UNTIL YOU COME BACK TO ME ON EARTH. MY BLESSINGS ON YOU. LYUDA AND VOLODYA AND ALL YOUR FRIENDS AND RELATIVES IN YAROSLAVL SEND YOU GREETINGS. UNTIL WE MEET AGAIN, MY DEAR ONE.

Valya produced what must have been the first tears shed in space, and they formed into perfect little *shariks* and floated gently into the air-conditioning filter.

"Dawn, please tell Mama not to worry," Valya transmitted.

On June 18 both Valya and Valery sent special messages to Nikita Sergeyevich, who was attending the Plenary Session of the Central Committee of the Communist Party of the Soviet Union in the Grand Palace of the Kremlin. They were read to the meeting by Leonid I. Brezhnev, chairman of the presidium of the Supreme Soviet.

Throughout the day Valya controlled the *Vostok 6* manually, using the stopwatch and the VZOR orientation device which looked down on earth. The spacecraft responded well to her touch, and she had no trouble in performing the maneuvers.

Since she had completed all the required work for her mission, she had time to perform "bonus" experiments for the doctors and engineers at the mission-control center.

In particular, they were interested in having her perform experiments to see whether weightlessness had yet disrupted the function of her vestibular apparatus. She continued to make drawings in her logbook of the spiral, star, and parallel lines and to write in it as well. These sketches were made on the twenty-ninth, thirty-first, thirty-third, and forty-fifth orbits for comparison with those made in the earlier orbits. However, all was not work. Valya had taken along with her photographs that her fellow cosmonauts and others at Star Town and Tyuratam had aked her to autograph while in orbit.

The first was "To my dearest Mama." Then she autographed pictures for Venera, Sergei Pavlovich, Andrian, the intense young spacecraft engineer, and the doctors. She also wrote a sentiment for Mstislav V. Keldysh, the chief theoretician of cosmonautics, who ranked with Sergei Pavlovich and with Valentin Petrovich Glushko as one of the three most important men in her country's space program.

Valya's last day in space ended with the fortieth orbit. Valery was just beginning his seventy-first orbit as she called to wish him good night.

She woke at 5:00 A.M., on June 19 and made a quick

report on the condition of the *sharik* and her own health. She hurriedly did her morning exercises, washed up, and ate breakfast. There were only a few orbits left before Valya would have to orient the spacecraft and prepare for the firing of the retrorocket that would brake it from orbit and send it flaming back into the earth's atmosphere.

"Sea Gull, this is Dawn," came Sergei Pavlovich's familiar voice over the radio. "Are you ready for reentry and landing?"

"This is Sea Gull," Valya replied. "I am all set."

"This is Hawk," Valery replied. "I'm all set, too."

"This is Dawn," said Sergei Pavlovich very firmly. "We are all ready here on earth. Everything will be done just as we all rehearsed it. Keep on your toes and go right by the plan."

As the forty-eighth orbit began, *Vostok 6* went into its last "night." Emerging from it, Valya firmly grasped the manual-control lever and oriented the spacecraft for the proper angle for reentry. Her nerves were steady and she went "right by the book," as Sergei Pavlovich had ordered.

She had time for a quick and final call to Valery. "Hawk, this is Sea Gull," she said. "I'll be waiting for you back on our green earth."

"Sea Gull, this is Hawk," Valery rejoined. "I won't keep you waiting, believe me!"

Now that the *Vostok 6* was traveling with its engine facing forward, Valya waited for the command to fire it.

"Sea Gull, this is Dawn," a voice crackled over the radio. "On the mark, fire retroengine. Three, two, one—mark!" The command came only thirty-four minutes and thirty-seven seconds after she had placed the ship in the proper orientation.

Valya flipped the retroengine "fire" switch.

Her weightlessness began to disappear, just as it had

appeared three days earlier while she was being inserted into orbit. She was pressed back into the ejection seat by the same "giant hand" that had done the same thing during lift-off. She glanced at the small globe on the instrument panel. She could tell that she would land as planned in the area of Karaganda, not far from where Andrian and Pavel had landed last year after the *Vostok 3* and *4* missions.

Then the *sharik* began oscillating slightly as it dipped into the denser air. The deceleration forces on Valya increased to 9 g's. Even turning her head was an effort. But she did, and looked out of the right porthole. As Yuri and Gherman had done before her, she elected not to close the blinds. There was no danger. The heat-resistant glass of the porthole could withstand the tremendous temperatures being generated by the friction of the air moving over the surface of the *sharik*.

At first the color outside the porthole was light rose, but it soon changed to scarlet, which in turn gave way to bright crimson. Molten metal from the radio antennas flowed like mercury over the windows. Finally the brilliance of the burning Vostok shut out even the sun through the left porthole. If it could have been seen from earth, *Vostok 6* would have looked like—and, in fact, was—a "shooting star."

Involuntarily, Valya glanced at the thermometer that recorded the temperature within the spacecraft. It showed 70° F. Outside, she knew, the temperature was a searing 18,000° F. It was an eerie situation. There she sat at a comfortable temperature in the midst of an inferno. It did not make sense. However, at no time was Valya afraid. Her training had fully prepared her for the rigors of reentry.

For sixteen minutes, Valya rode the *Vostok 6* through its blazing reentry. Then, at an altitude of little more than four miles, explosive bolts blew the hatch off the *sharik*.

Exactly two seconds later, small rockets attached to the ejection seat fired and blasted Valya out of the ship. A barometer fuse automatically opened the parachute, and Valya felt the familiar jerk as the white canopy with red concentric circles filled with air and checked her fall toward the ground below.

Below her Valya could see a river and a lake. For a moment or so, it looked as if she might land in water. She was not worried, however; she had parachuted into water before during training. Also, she had a dinghy that would automatically inflate and support her if she should land in the lake or the river. The wind shifted slightly, and she took advantage of it to guide the parachute to a small clearing among a grove of birches. She was some 384 miles northeast of Karaganda, a large industrial city in Kazakhstan. The official records of the International Astronautical Federation put the point more precisely: 53° 56′ 18″ north and 80° 27′ 34″ east. She had traveled 1,222,014 miles to reach that spot.

The time was 11:20 A.M., and Valya had landed after seventy hours and fifty minutes in space. Her *sharik* had landed five minutes earlier, and she wanted to examine it as soon as possible. She hurried over to it, and by the time she arrived, there was already a crowd around it. Valya had been spotted in the air by two men working on a bridge. They had spread the word to the nearby collective farm, and workers came running to meet her. At almost the same time, a small airplane appeared overhead and several parachutes blossomed from it. Naturally, it was a medical team. A doctor quickly shook off his parachute and rushed for Valya with the inevitable syringe. She had not been on earth for five minutes and these vampires wanted her blood!

"What has happened to your nose?" one of the medical personnel asked suspiciously.

"I bumped it while I was landing," she said. "It's nothing serious. I'll live."

In keeping with an old Russian custom, the men from the neighboring collective farms welcomed Valya with the traditional gift of black bread and salt. But more appetizing after three days in space for a former country girl were the cheese, *lepeshki* (griddle cakes), and *kumiss* (fermented mare's milk) they also brought her. Space food was tasty enough in orbit, but Valya was back on earth and hungry!

Almost before she knew it, Valya was hoisted into a helicopter and flown off for a rendezvous with Valery, who had landed at 2:06 P.M. at a point about 330 miles northeast of Karaganda. The two met in a small village on the banks of the Volga River. Valery was already there. Like Valya, he had by then changed into a two-piece blue gym suit. There

Back on earth, Valya's first task was to check her spacecraft.

was a difference, however, between the two newest space travelers. One was heavily bearded. Valya rushed to Valery and hugged him as hard as she could. Then she kissed him.

"Take it easy, Valya," he said with a wink. "It's I, Valery, not Andrian."

Valya blushed.

Poor Valya! She had not realized that lovers are the worst security risks in the world. She could keep the fact from her mother and family that she was going into space; that was easy. There, she was dealing with written communications. But one simply cannot meet a special man every day in the week and act as though he were the postman, the milkman, or the cabdriver. Her fellow cosmonauts saw through her pretense all too easily. Valya was a great cosmonaut trainee but the world's worst actress. The same evaluation applied to Andrian. He was a great cosmonaut but an absolutely horrible actor. The fact that Valya and Andrian were in love well before the *Vostok 5* and *6* mission was apparent to everyone in Star Town.

When the wives of the cosmonauts got together, Valya was always the first topic for the day's gossip. "When will she join us?" Discussion was generally based on a time scale. About half the wives thought she would marry Andrian before they went into space. The other half thought they would be married immediately after Valya's mission. Both suppositions, as later events proved, were wrong.

A very delicate situation had arisen. The Soviet government realized that a marriage so soon after the mission could raise embarrassing questions. Some people might think that the whole purpose of the mission had been merely to see whether space flight would affect the ability of a man and woman to produce a healthy baby. Valya was told to deny that she was engaged and that she was going to marry anyone.

Hawk and Sea Gull are welcomed by their colleagues.

Yuri and Andrian, several other cosmonauts, and several reporters were among the party that greeted Valya and Valery in the little village. Sergei Pavlovich was with them.

"Well, Valya," he said, "you have proved that Russian women are the equal to Russian men in space, and there was never any doubt about the equality on earth!"

That evening Valya and Andrian took a walk along the banks of the Volga. This time when their hands clasped, Andrian did not draw away. Instead he tightened his in an affectionate squeeze.

"Did you get my message from space?" Valya asked.

"Yes, I did," he answered. "And so did everyone else. Half of them thought it was to your mother and the other half thought it was to a boyfriend in Yaroslavl. I guess our secret is still safe."

What Valya was alluding to was a communication she had sent to Andrian early in her mission: "Here's a message to the dearest human being on earth," she had radioed mysteriously to the mission-control center.

Now, the moon arose across the Volga, its image in the water formed a silvery bridge from bank to bank.

"That's our bridge to the future," Valya said pensively to Andrian.

The welcome-back in Moscow on June 22 was completely beyond Valya's imagination or expectations, even though she had seen the receptions for Andrian and Pavel.

The Il-18 carrying Valya and Valery rolled to a stop several hundred feet from the flower-covered reviewing stand which had been erected just outside the terminal. They remained inside the plane until the traditional red carpet had been adjusted at the foot of the steps and across the concrete still wet from a recent shower.

Then, as the army band struck up the air force song "Higher and Higher," the two cosmonauts began striding down the "one hundred steps to glory." Unconsciously, Valya reached out and took Valery's hand. Thus there was another first for the Soviet space program. For the first time, two cosmonauts walked hand in hand to report to the premier of the Soviet Union that they had completed their space mission!

On the stand were the usual governmental figures and party functionaries. Yuri and Andrian were there, too. Almost lost among them were Mama, Volodya, and Lyuda.

After making their formal report to Khrushchev, Valya and Valery stood at attention while the band began playing "The Internationale," the national anthem of the Soviet

Union. Never the one for formality, Nikita Sergeyevich committed a serious breach of protocol when he impulsively gave Valya a bear-hug and kiss before the national anthem was finished.

After the ceremonies at the airport were finished, Valya and Valery climbed into the back of an open car for the trip to Red Square. Lenin Prospekt, down which the parade moved, was thronged with happy Muscovites, screaming Valya's name and waving huge pictures of her. Everyone seemed to have forgotten that Valery had been in space, too. Dimitrov Street also was packed as the car moved slowly along it and then across the bridge over the Moscow River and into Red Square.

The enthusiasm that filled Red Square simply overwhelmed Valya. She stood on top of Lenin's Tomb with Nikita Sergeyevich, who could not resist another bear-hug and kiss. With them stood Yuri, Andrian, Gherman, and Valery.

Since she was several pounds lighter after her mission, her navy blue dress fitted her a little loosely. The white high-heeled shoes were certainly more feminine than the space boots she had worn shortly before this event. Her recently touched-up kitten coiffure stood up well under the breeze blowing across the square.

Below the group on top of the tomb, in absolute delirium surged a crowd of proud Muscovites. Some carried large pictures of Valya and Valery. Others carried banners and signs. Large groups of students chanted their names in unison. Valya and Valery waved. Carried away with the excitement, Valya clasped her hands above her head like a boxer and shook them. Suddenly remembering where she was and with whom she was standing, she composed herself and acted more reserved, except for a grin from ear to ear.

Nikita Sergeyevich was not known for his decorum. He

Along the Lenin Prospekt, portraits of Valya, Valery, and Lenin. The posters say "Victory!" "First in the world, a Russian woman cosmonaut!" and "Thanks to the Communist Party of the Soviet Union!"

could not help injecting a note of anti-Americanism into the celebration.

Reading very gravely from a speech, he duly noted the American contributions to manned space exploration. "We welcome the space flights of the American astronauts. This is a noble and worthy partnership in the competition for a peaceful conquest of outer space," he said rather grandly.

Then, with a gleam in his eyes and a grin like a happy troll, he put aside his written speech. "Bourgeois society always emphasizes that woman is the weaker sex," he stated. "That is not so. Our Russian woman showed the American astronauts a thing or two. Her mission was longer than that of all the Americans put together!"

With that remark, he turned to Valya, very deliberately pointed his finger at her, and paused. "There is your weaker sex!" he said.

Nikita Sergeyevich, the politician, went on to use the occasion to rattle a space saber. He told the crowd that rockets capable of launching men into orbit about earth could also be used for other purposes.

As Nikita Sergeyevich spoke, only a few in the crowd

Most of Moscow crowded into Red Square to hail the world's first woman in space.

below Lenin's Tomb noticed a slight commotion among the cosmonauts who looked down at them, waving and smiling.

The cosmonauts had elbowed, pushed, and tugged Andrian until he was next to Valya. She reached down and clasped his hand and squeezed. Again, Andrian did not jerk it away.

"Let's tell Nikita Sergeyevich," she whispered to Andrian.

"Don't you even think about it," he said sternly. "Sergei Pavlovich said no one, and he outranks Nikita Sergeyevich!" While Premier Khrushchev might have disagreed, Sergei Pavlovich had no doubts. Valya did not say a word to Nikita Sergeyevich.

No sooner had Valya and Valery returned to Star Town than the reporters from the Soviet press began besieging General Kamanin's office with requests for interviews.

Valentin Goltsev, a reporter from *Izvestia,* had an original idea for getting to see Valya. He showed up in the general's office with a copy of a book, *Woman into Space,* by Jeri Cobb, the American woman who had so much wanted to be an astronaut, only to be turned down by the National Aeronautics and Space Administration. He briefly described the book to General Kamanin and asked if it would be possible to take it to Valya.

"Well, it seems to be a sad story. But I guess we can find the time to show it to Valentina Tereshkova. She probably will be interested in it," he said.

General Kamanin accompanied reporter Goltsev from his office to the hospital, where Valery and Valya had been taken after the reception in Red Square. They took the elevator up to the second floor, and Goltsev followed General Kamanin down the corridor. Suddenly the general stopped before a door and told the reporter to wait there. He went in and stayed for a few minutes. Then he came out, accompanied by Valya and Valery. She had on a light blue blouse and a pair of very tight fitting black slacks.

"We are happy to have guests. The doctors are very strict and do not allow even our relatives to visit us," she said to Goltsev.

Valya invited the visitors into the room. On the table next to her bed were two bowls containing black currants and gooseberries.

"Help yourself to natural vitamins! Good for the health. And you don't have to be a cosmonaut to benefit from them!" she said.

The reporter handed Valya the book. She looked at it in puzzlement because she did not know English.

"It's about you, Valya. The Americans have published this book about you," said General Kamanin with a straight face, Valery, who knew English, saw at a glance what it was, but he went along with the joke.

However, neither of them was a very convincing actor. Valya knew something was up.

"This is a book by Jeri Cobb," said Valery. "I have heard of her and I really sympathize with her. She is a daring and courageous woman. It's a disgrace that the Americans have treated her as they have," he said.

Valya grinned, and quickly grabbed the book and began thumbing through it. She was especially interested in its many pictures. Valery and General Kamanin translated the captions for her.

"I really pity Jeri Cobb. She is an excellent pilot and a brave woman. It's not easy to fly a jet plane; it requires great strength, good training, and quick reflexes. When I think of Jeri's failure to become a cosmonaut, I know that it is not a personal thing," she said.

On the evening of June 22, there was a state reception for Valery and Valya in the Grand Palace of the Kremlin.

A quiet moment in the midst of celebrations: Valya with Sergei Pavlovich Korolev, Yuri Gagarin, and Valery Bykovsky.

Among the hundreds of celebrities and dignitaries who milled among the white marble columns of St. George's Hall and under the ornate gilt ceiling of the Hall of Mirrors were lithe and delicate ballerinas, beautiful movie actresses and handsome actors, grim, be-medaled generals and admirals, serious-faced composers, and foreign diplomats in striped trousers.

All these rubbed elbows with a collection of some of Sergei Pavlovich's designers and engineers, who found themselves acutely out of place among famous circus stars, high party officials, famous authors, and the inevitable newspaper reporters. Nikita Sergeyevich was there, too, beaming and bear-hugging. Valya and Valery felt equally out of place, even though they glimpsed, now and then, familiar faces such as Yuri, Gherman, Pavel, their wives, General

Kamanin, and of course, Sergei Pavlovich and his old friend Valentin Glushko. From time to time, Valya managed to wave to Mama, Lyuda, and Volodya, huddled together in bewilderment and awe.

During the formal aspects of the reception, Leonid I. Brezhnev, chairman of the presidium of the Supreme Soviet, bestowed on Valya and Valery the rating Pilot-Cosmonaut and the title Hero of the Soviet Union, the highest honor that can be given to a Soviet citizen. He also presented them with the Order of Lenin, and the Gold Star which goes with the title Hero of the Soviet Union. As if that were not enough, he beamingly announced to the guests that a bronze bust of Valya would join that of Yuri along the Cosmonaut Way that leads to the Space Monument in Moscow.

Following the awards, the band struck up a quick march, the traditional signal for the toasts that are an immutable part of any Russian social function. Nikita Sergeyevich led off, touching his small glass of vodka to those of Valya and Valery, in turn. Soon the toasts were coming from every direction, each accompanied by a more glowing and, to Valya and Valery, more embarrassing account of their deeds in space.

It was all too much for the tomboy from Yaroslavl. As soon as possible, when the guests began toasting each other, she, Mama, Lyuda, and Volodya slipped away to the Bolshoi Theater to see *Carmen*. Even though they waited until the houselights were down before entering their box, the word was out that Valya was there. One brash American journalist even invaded the private box for an interview.

Following Valya's flight, skepticism was voiced in some Western newspapers. They pointed out that it was not really much of a feat if it had been accomplished by a former loom operator.

Gherman Titov immediately came to Valya's defense. In the October issue of *Aviatisiya i Kosmonavtika*, a journal of which he would later become deputy editor, he wrote:

> But they forget that Valya boarded the cabin of *Vostok 6* not from a weaver's loom but from the controls of an airplane. Valya was a first-class parachutist. She went through flight training with us on modern planes, trained at various test installations, and prepared for the flight physically and mentally. She went step by step over the long, entire, and difficult route that cosmonauts must travel. This must not be underestimated.

Wedding Bells for Sea Gull

Letters by the thousands began arriving at Star Town—all addressed to Valya. They were from lonely bachelors from every corner of the USSR, and almost every one proposed marriage! Valya could have had her choice of Russian men.

Her quick grin, trim figure, and ash-blond hair made her something of a sex symbol for the puritanical press and television of the Soviet Union. The sight of Valya even brought a plaintive and melancholy sigh from Mikhail Sholokov, the Nobel Prize winner and famous author of *And Quiet Flows the Don*. He said to a reporter:

> My age and somewhat conservative makeup compelled me to believe we men were arbiters of the mind —soldiers and salt of the earth. But what do we see now? A woman in space. Say what you will, but this is incomprehensible. It contradicts all my set conceptions of the world and its possibilities. She'll now have proposals of hand and heart showered on her by the thousands. Alas, I who have carried the cross of married life these forty years can offer her neither heart nor hand, but I can embrace her and wish her all the best.

The Kremlin reception of June 22 was only the first of many official visits and appearances that Valya would make. Her first subsequent public appearance took place at an event that almost turned into a complete shambles. Her *Vostok 6* mission had been scheduled to take place just before the World Women's Congress on June 24 in the newly opened glass-walled Palace of Congresses in the Kremlin. On that day, two thousand women from more than one hundred countries gathered at the call of the Women's International Democratic Federation, an organization composed largely of Communist women.

Premier Khrushchev was on hand to open the congress with a denunciation of the United States as preparing for war "on a scope unparalleled in history." Having done his duty to the party, he introduced Valya to the women. Immediately Chairwoman Ezheni Kotton hugged and kissed Valya as the delegates left their seats and crowded around the podium cheering and reaching for her.

After pointing out that Valya had lost her father in World War II, Nikita Sergeyevich beat the drum for communism. "She stood firmly on her feet and now has risen to such heights as no person in the capitalist world brought up by wealthy daddies and mommies can rise," he boasted.

Valery and Valya were honored guests at the World Women's Congress in Moscow on June 24, 1963.

There was much partisan dispute among the delegates from different countries, who were unable to agree on what should be the aims of the congress, but they were united in one thing, their admiration and approval of Valya, who on the second day made a short speech, carefully refraining from anything that might add fuel to the explosive atmosphere. Literally, there was not an empty seat among the six thousand in the palace.

"Honored ladies! Fellow delegates! Dear friends," she said in a completely unnatural voice as she tried desperately to remember the speech she had memorized only a short time previously. "The Soviet people have placed a great deal of confidence in me, a simple Russian woman. They entrusted me with a flight into space. I am happy that I was able to complete the programmed flight of the *Vostok 6* and did it completely."

She went on to talk about her mission in more or less technical details, with material right out of *Pravda*. However, the audience did not seem to mind the repetition.

"From space I sent radio messages, greetings, to the people of every continent over which *Vostok 6* orbited," she said in closing. "Today, I personally, and with great sincerity and friendship, greet all you women of our planet who have gathered here in my motherland for this meeting." When she was finished, the women mobbed her as she stepped from the platform. All of them wanted to touch her, get an autograph, or merely speak to her.

The congress ended with the adoption of a statement which appealed to women of the world to fight for peace and to work against the arms race.

Valya's next public appearance was at a press conference held on June 25 at Moscow State University. The conference was opened by Dr. Mstislav V. Keldysh, president of the Academy of Sciences of the USSR and also chief theoretician of cosmonautics. He announced that the academy had decided to award Valya and Valery yet another honor, the prestigious Tsiolkovsky Gold Medal for outstanding achievement in cosmonautics. He was followed with remarks by the head of the academy's Department of Technical Sciences. On hand to lend moral support were Yuri, Gherman, Pavel, and Andrian.

The reporters squirmed through the talks on the technical details of the *Vostok 5* and *6* missions. However, they came to life when Valery got up to speak. He briefly told them of his mission and how he had responded during it. There were a few questions from the correspondents. Among them was one that gave Valery a chance to express his attitude toward press conferences. A commentator from the Soviet radio broadcasting service asked him which planet he would like to visit first.

"To begin with I'd like to try the moon, and then Mars and Venus. . . . My health is excellent, and I feel fine. All I need is a little rest after such interviews as this one," he said shortly.

However, it was evident that Valya was going to be the star attraction, even though she appeared last on the program. When she stood up to read her prepared statement, the journalists broke into prolonged applause, which both pleased and embarrassed her. However, she managed her spontaneous and by then famous grin.

During her question and answer session, she received queries from both Soviet and Western reporters. She answered them quickly and with a minimum of polemics.

"Were you afraid at any time during the launch?" asked the reporter from *Newsweek* magazine.

"In Russia we have a saying 'If you are afraid of wolves, don't go in the woods,' " she said with pride. "What was there to be afraid of? I didn't doubt for a minute the reliability of my spacecraft."

Several more questions followed on various aspects of her mission, and then the newsmen got down to something that really interested them.

"Who is the person dearest to you on earth?" asked a reporter who had heard her ambiguous message from *Vostok 6*.

"That's easy." Valya spoke quickly but suddenly hesitated. "It's—my mother."

Everyone laughed at her sudden blush and quick grin. The rumors about her and Andrian had been all over Moscow for weeks.

After the tedious press conference and official ceremonies were over, Valya found time to visit Yaroslavl, only to encounter another tumultuous welcome, beginning with a boat ride up the Volga. She was not nearly as excited telling

Yelena Fedorovna about her *Vostok 6* mission as she was in breaking the news that she and Andrian were to be married in several months. She told her while strolling in back of the house among the birches. First, she explained, she had to make a world tour with Yuri and Valery; then she and Andrian would be married in early November.

"Perhaps," she said lightly, "I can get Nikita Sergeyevich to sit as sponsor with you and Andrian's mother, since neither he nor I have a father."

Her mother looked at her as if space travel had affected her brain.

Once the most exciting news was over, Valya proceeded to tell her family and friends about her space flight. It was a story that she was to repeat many times in the following days. She told it at the tire factory, and many times in many places at Red Perekop. At a special rally of the Komsomol, the flowers for Valya covered the speaker's table and literally overflowed on the floor.

In the mill, she tried her hand at the loom again for old times' sake. But several years of astronaut training only proved that she had lost her skill as a weaver! She also found time to take a swim in both the Volga and the Kotorosl, but she did not attempt to dive off the Kotorosl bridge. She did visit the Yaroslavl Air Sports Club, however, and she made a couple of jumps just to show her old friends that she had not forgotten what she had learned with them. She also explained in detail the jump she made during her *Vostok 6* mission.

Valya's trip to Maslennikovo was even more nostalgic. Suddenly the two years in the hustle of Moscow seemed to vanish. In the quiet of the small hamlet time seemed to have stood still. Except for one thing. The hut in which she had been born was the scene of a lot of noise and carpentry. It was being converted into a library for the collective farmers in the area.

Just as suddenly, Valya was brought back to reality. She returned to Moscow for a world tour of some of the thirty countries which had officially invited her.

It began with a trip to Czechoslovakia with Yuri on August 14. The big Tu-124 landed at the Ruzyne Airport in Prague at 9 A.M. In what was to become a familiar procedure, Valya climbed into an open limousine and headed for a strenuous round of activities. The car moved slowly across the Moldau River and through throngs of cheering people, waving signs proclaiming *Buda Vitana, Valentina*! ("Welcome, Valentina!") and *At Zije Valentina* ("Long live Valentina!").

She and Yuri appeared on television in Prague. They were guests of the Academy of Sciences, and President Antonin Novotny awarded Valya the title of Hero of Socialist Labor of Czechoslovakia. Later she met and talked with Premier Alexander Dubcek.

In the succeeding days she and Yuri visited Bratislava, Lidice, and Ostrava. In Bratislava, she visited a textile plant that reminded her of Red Perekop, and she told the workers there how much at home she felt. In addition, she and Yuri were made honorary citizens of the city.

On August 21 the two left Prague for Moscow, with Valya bearing, among other gifts, a doll dressed in Czech national costume and a small puppy.

The next stop, a month later, was in Varna, Bulgaria, where she was greeted by Todor Zhivkov, chairman of the Council of Ministers of Bulgaria. Here she was joined by Valery, who was to make part of the tour with her.

After a brief rest in Moscow, Valya again took to the road on September 30. She had been invited to visit Cuba by Premier Fidel Castro. This time she flew in a Tu-114 Aeroflot passenger plane, but she left Yuri and Valery behind. On October 1 she landed in Havana. Shading her eyes against the tropical sun, she walked down the steps wearing a dark blue skirt, red blouse, and her white high-heeled shoes, a color combination that matched the many Cuban flags fluttering about the airport.

She was met at the airport by President Osvaldo Dorticós Torrado and his wife; Premier Castro; Soviet Ambassador Alexander Alexeev; Señora Vilma Espin, president of the Federation of Cuban Women; and Cuban protocol chief Antonio Carillo. Then came a motorcade through Havana, with streets packed with people waving Soviet and Cuban flags. Valya and the dignitaries rode in a convertible that had been given to Castro by Premier Khrushchev.

On the following day, she had something of a holiday at the resort beach of Playa Giron, some 125 miles southeast of Havana on the Bay of Pigs. There was time for strolling on the beach and a boat ride, all of which she duly recorded with a movie camera to show friends back in Star Town.

The following day began the inevitable round of visits and ceremonies. There was a visit and wreath of flowers for the statue of José Martí, the Spanish poet and fighter for Cuban freedom; a visit to La Cabaña Castle; and a reception by Raul Castro, the defense minister, who presented her with some models of a ship and an airplane. She also received the Alas de Piloto de Combate (pilot's wings) from the Cuban air force. That evening she appeared on television and answered questions from Cuban journalists.

Among the interesting things she revealed was the fact that Yuri Gagarin had been appointed commander of the first lunar landing mission and that she was a member of it. Then, diplomatically, she added that she was going to propose that a Cuban woman be included on the mission.

The pace of activities increased on the next day. She visited the crew of the Soviet freighter *Valentina Teresh-*

kova at dock. Then there was a visit to Colina University and a tour of the Ariguanabo Textile Factory.

That night she made an appearance at the Chaplin Theater in Havana, where she spoke to an enthusiastic crowd. Extra buses had to be added to the service to accommodate the crowd attending the theater.

Time passed rather quickly in a whirl of social and diplomatic appearances. On October 10 Valya returned to the airport with Premier Castro to welcome Yuri and his wife, Valentina, who were to accompany her to Mexico. That afternoon she engaged in charitable work, assisting Cuban relief authorities in sorting clothes that had been collected for the victims of Hurricane Flora, which had earlier taken some one thousand lives when it hit the western end of the island.

By this time, Valya was frankly beginning to tire of the heat, humidity, and sugar cane fields as well as Premier Castro's Latin graciousness and charm. In addition, her mind was on a much more important date only a month away.

On October 11 Yuri and his Valentina and Valya flew to Mexico City. They had been invited to visit the country by the Club Aero Mexicano and the officials of the Fifty-sixth General Conference of the International Aeronautical Federation, which was meeting there.

After an enthusiastic reception at the airport and a ride into the city in an open limousine, the first item on the agenda was, of course, a press conference. The reporters were inquisitive as usual, and Valya had the right answers.

"What major influence do you feel your mission has had upon women of the world?" one asked.

"Since 1917 Soviet women have had the same prerogatives and rights as men," she said primly. "They share the same tasks. They are workers, navigators, chemists, aviators, engineers. And now the nation has selected me for the honor of being a cosmonaut. As you can see, on earth, at sea, and in the sky, Soviet women are the equal of men."

By her side Valya Gagarina nodded in assent.

From the press conference, Valya and Yuri rushed to the hotel to change before a meeting that evening with President López Mateos in Los Pinos. Yuri presented the president with six autographed pictures of cosmonauts and a plastic model of the Vostok. Valya gave him six small plastic Vostoks. Each one had a picture of the cosmonaut who had flown it embedded in it.

On the following day Valya and Yuri and their party visited the University of Mexico, where they mingled with the students, signing autographs and answering many questions concerning their Vostok missions. There also was a press conference in which most of the questions centered around Soviet moon-landing plans. Both dodged the questions adroitly.

October 13 found the travelers at a 9 A.M. ballet at the Palacio de Bellas Artes and then driving off to view the ruins of Teotihuacán. Since both of them were in top physical shape, the climb up the Pyramid of the Sun was no strain. Less athletic members of the party prudently remained at the base. That evening there was a cocktail party in honor of Valya at the Hotel El Diplomatico.

On the following day President Mateos opened the conference of the International Aeronautical Federation. When Yuri and Valya were introduced, delegates from fifty-nine countries arose and applauded loud and long regardless of their political philosophy or their nations' diplomatic relations with the USSR. Valya sat next to the American woman pilot, Jeri Cobb, whose book the reporter from *Izvestia* had given her on her return from space.

That evening there was a formal banquet. Almost the entire diplomatic corps of Mexico City turned out. While the American ambassador did not appear, the military attachés of the embassy were present.

Weary and sun-tanned, Valya and her party embarked upon their Il-18, with three American navigators, for the flight to New York on October 15. The plane landed in New York a little after 9 P.M. Valya and Yuri emerged to be met by a milling, shoving crowd of some thirty or forty newspaper photographers and a beaming Ambassador Anatoly Dobrynin.

Dobrynin announced firmly that the USSR was embarked on a manned lunar landing program and that it would culminate before 1970. He then looked toward Yuri, who said simply, "Yes, of course!"

Since early October the rumors had been flying about Moscow that Valya and Andrian were engaged and that they would be married as soon as she returned to the Soviet Union. However, Valya was still under orders to deny everything. A reporter asked about the rumors. Valya said, "You will certainly know when I do get married, because it is impossible to hide anything from the press anywhere."

She then volunteered the information that even if she did get married it would not hinder her career as a cosmonaut, since many of her fellow cosmonauts, male and female, were married.

Another reporter asked her if she would prefer to be married in space or on earth.

"If that should happen during a space flight," she said, "there would be plenty of room to get married in—and most important, there will be no newspaper reporters."

Valya and Yuri then entered a car and drove directly to the United Nations building. This time, the streets were not lined with the crowds to which they had been accustomed. Their visit was not an official one, and no preparations had been made for it.

They were greeted at the UN by Secretary-General U Thant, who introduced them to the General Assembly. Its members gave them a standing ovation. After their brief greeting to the body, U Thant presented them each with a packet of UN stamps dedicated to the peaceful uses of outer space.

Their schedule permitted only a brief tour of New York, but Yuri found time to buy a kiddie car for his four-year-old daughter Yelena. Valya contented herself with a brief window-shopping tour along Fifth Avenue. For a while both watched the ice skaters in Rockefeller Plaza, marveling at a skating rink sunken in the concrete canyon of midtown Manhattan.

That night Yuri and Valya boarded their Il-18 turboprop for East Germany. A cheering crowd greeted them at Schoenfeld, in East Berlin, when the plane rolled to a stop. For three days they received similar welcomes as they toured the city, seeing the sights, such as the Berlin Wall from a vantage point near the Brandenburg Gate. Valya was interviewed on television, and she revealed that some of the female cosmonauts training with her were married and had children. Then came a few days in Poland, where she and Valery were met at the plane by President Wladyslaw Gomulka. Again she was on television, and she told an admiring Polish audience that other women would follow her into space.

"They will go up in spacecraft with mixed crews so the men will not get bored so quickly," she said on an uncharacteristically racy note.

After Poland, Valya returned to Moscow for the wedding that everyone was anticipating, despite the fact that it was supposed to be a secret.

The wedding of Valentina Vladimirovna Tereshkova and Andrian Grigoryevich Nikolayev was undoubtedly the greatest in Moscow since Czar Nicholas II had married Princess Alexandra of Hesse-Darmstadt, Queen Victoria's granddaughter, in 1894.

It was a bone-chilling 30° F. on the sunny morning of November 3, 1963. Despite the cold and the fact that Valya's wedding was still officially a secret, more than a thousand milling and excited Muscovites crowded around the Marriage Palace on Griboyedov Street. The palace had been opened in 1961 in an attempt to entice romantic young Muscovites away from having weddings in churches, a practice frowned upon by the state. Now it was all the blue-uniformed *militzia* could do to keep the crowd around the palace behind their cordon.

Just before noon, a long black Zis limousine drew up in front of the palace, and Valya got out, wearing a gray fur coat with the collar turned up around her ears. Despite the cold she was not wearing a hat. Andrian got out next, wearing a dark overcoat but no hat. Two other cars pulled up behind them, and Mama, Lyuda, and Volodya got out of one. Out of the other came Anna Alekseyevna, Andrian's mother; Ivan and Pyotr, his brothers; and Zinaida, his sister. Behind them in yet another car came Yuri Gagarin and his Valya, Gherman Titov and Tamara, and Valery Bykovsky and his Valya. Valery remarked to his wife that this was their second trip to the Marriage Palace.

In the crowd being held back by the *militzia* were several disappointed couples who had showed up at the Marriage Palace for their wedding only to be turned away temporarily!

"What's going on?" one anguished groom yelled. "We were supposed to be married at this time! What's happening here?"

"Come back at one o'clock, and we'll take care of you," said one of the Marriage Palace officials.

When the crowd saw Valya step from the car, it set up a chant of "Valya, Valya, Valya." Valya waved and flashed her familiar grin but quickly ran into the Marriage Palace.

Once inside, Valya and Andrian removed their overcoats. Valya was wearing a white, low-necked knee-length dress with a veil caught off her face and elbow-length gloves. Andrian was dressed in a black suit, but he wore a silver necktie for the occasion. The loudspeakers in the corners of the room above the wooden paneling suddenly blared into Tchaikovsky's First Piano Concerto as Valya and Andrian walked together down the worn carpet to a large but very plain walnut table. In the center of the table was a large book bound in red leather. Seated behind the table were

The wedding of Sea Gull and Falcon, November 3, 1963.

Yelena Fedorovna and Anna Alekseyevna with the beaming surrogate father of Valya and Andrian, Nikita Sergeyevich Khrushchev.

The ceremony was brief. Mrs. Yevdokiya G. Voroshilova, the registrar of marriages, after stressing the importance of the day, got down to business.

"In conformance with the law of marriage and the family of the Russian Soviet Federated Socialist Republic, the marriage of the citizens of the Soviet Union, Heroes of the Soviet Union, Pilot-Cosmonauts Valentina Vladimirovna Tereshkova and Andrian Grigoryevich Nikolayev is being registered."

Valya and Andrian stepped up to the registry book and signed their names. For the first time Valya signed her new name, Valentina Vladimirovna Nikolayeva-Tereshkova. After Andrian had signed his name, Valentina Gagarina signed as witness for the bride, and Valery Bykovsky signed as witness for the groom. Thus Valya and Andrian became marriage no. 10,860 in Mrs. Voroshilova's register.

Andrian then slipped a ring on the third finger of Valya's right hand, where wedding rings are customarily worn in Europe. When it came Valya's time to place a ring on Andrian's finger, it took her almost a minute of nervous fumbling and pushing.

"Valentina Vladimirovna, you will now kiss your husband," Mrs. Voroshilova ordered, unnecessarily as it turned out. With tears in her eyes, Valya did not need a second command before complying.

Suddenly the room exploded with flash bulbs. Photographers who had been clustered in front of the table began flashing bulbs and switching on klieg lights as movie and television cameras began to hum. Vladimir F. Promyslov, the mayor of Moscow, stepped forward, and after making a short speech, handed the marriage certificate to Andrian.

Several other functionaries added their congratulations. Then a group of party officials began surging around the pair shouting their best wishes.

"So finally Falcon has married Sea Gull," said General Kamanin as he congratulated his cosmonauts.

In the crush of the crowd of government officials, party functionaries, and fellow cosmonauts and their wives, Yelena Fedorovna and Anna Alekseyevna had gotten pushed farther and farther back into the crowd around the couple. As a result, they were among the last to reach their children. Valya's eyes were already wet, but the sight of Mama approaching her brought on a new flood of tears. Even Andrian's eyes were moistened as he embraced his mother. Lyuda was crying, too, of course, as she hugged Valya. Volodya grinned in embarrassment as he shook hands with his famous new brother-in-law.

After a toast in champagne by Nikita Sergeyevich, the couple left the Marriage Palace and returned to their limousine for the trip to the reception. It took them to the southwest part of Moscow and the Lenin Hills, near the State University of Moscow, which looked more than ever like a huge wedding cake. The reception was in a government guest house usually used for visiting dignitaries from foreign countries.

More than three hundred guests were waiting for them, and they burst into applause as Valya and Andrian entered arm-in-arm with Premier Khrushchev's wife, Nina Petrovna, followed by her beaming husband, who was escorting the mothers of the bride and groom. It was to be a wedding reception such as Moscow had seldom seen.

The agenda called for a very formal and proper wedding reception. However, the personalities and heritage of the people involved made such formality and propriety impossible. It turned out to be a real, old-time Russian wedding reception, peasant-style.

With Sergei Pavlovich at the wedding reception.

The bride and groom and Nikita Sergeyevich and Nina Petrovna sat at the head table with other dignitaries and cosmonauts, dining on everything from caviar and soup to fish and ice cream. The solemnity of the occasion fell apart when Valya, Andrian, and Nikita Sergeyevich picked up their glasses of champagne and cognac and began hopping from table to table. The signal for this departure from protocol began when Nikita Sergeyevich rang a bell and proposed the first of literally hundreds of toasts that were to follow, including twenty-one proposed by himself!

The first toast was to the couple's happiness. "I wish you further success in the air and space," he said simply.

As Valya and Andrian moved among the tables, they were followed by Nikita Sergeyevich, who became happier and happier after stopping at each table. In one hand he carried a bouquet that Valya had given him earlier, and in the other a glass, not often empty. Typical of the groups at which the three stopped was a table of actors and singers from the Bolshoi Theater, who promptly began singing a song they had made up for the occasion. As the four-hour reception continued, their voices, not always in the harmony they exhibited on stage, provided a counterpoint to the growing number of toasts that were being given to the bride and groom from the head table.

As the reception continued, from time to time Nikita Sergeyevich would ring a bell, bringing the festivities to a sudden halt. When the noise died down, he would yell "*Gorko! Gorko!*" ("Bitter! Bitter!"). According to an old Russian custom, this meant that the wedding wine had become bitter and could be sweetened only by a hug and a kiss of the newly wedded couple. Valya and Andrian obligingly made the wine sweet again.

In between his *gorkos*, Nikita Sergeyevich revealed some interesting psychological insights put forth in mixed metaphors.

"May I wish you long life. May the marriage be a long one. You have already reached great moral and cosmic heights," he said a little shakily.

After a brief period of introspection as a long-married man, he continued, "May you have radar to avoid all the griefs and obstacles of life"—being careful not to look in the direction of Nina Petrovna.

Somewhat later, he added a note that is common sooner or later at all wedding receptions: "If you have a baby, the gifts will come from all over the world." At this point Defense Minister Rodion Y. Malinovsky and Air Force Commander K. A. Vershinin called Valya and Andrian forward and with mock formality presented them with a baby doll that had a pacifier in its mouth. It was a *symbolic* gift, they said, grinning.

The reception went on for several hours. As the toasts became more frequent the guests began a variety of dances depending upon their ethnic or regional backgrounds. By all standards, it was the biggest and in some ways most successful wedding reception since the days of the czars.

After the reception, of course, came the honeymoon. With typical Soviet pragmatism, the government decided to combine it with a good-will tour. In other words, Valya would take up her world tour but would now have Andrian along as well as Valery and his wife, Valentina.

The honeymoon was a month's visit to India, Nepal, Ceylon, Indonesia, and Burma.

On November 9 the quartet left Moscow, and they landed the next day in New Delhi. Valya was the first off the plane, looking very feminine in her brown print dress and upswept hairdo. The party was met by Mrs. Indira Gandhi, the niece of the prime minister; Mrs. Nurudin Ahmed, wife of the mayor of New Delhi; General Georglyad, deputy director of the Ministry of Foreign Affairs; Soviet Ambassador I. A. Benediktov; several delegations of Indian women; and a delirious crowd of fifteen thousand.

Following an old Indian custom, Ambassador Benediktov welcomed his newly-wed Russian comrades with the wish that "every day of their lives would equal one hundred days of happiness." However, he pointedly refrained from following up with the other traditional wish that the couple have one hundred children!

The first stop on what must have been the busiest honeymoon in history was the Soviet embassy in New Delhi for a press conference which Prime Minister Jawaharlal Nehru attended. More than three hundred people milled about shouting questions in English and Russian. The one question repeated above others was whether Valya had thought about marriage when she was in space.

"We don't want to reveal all our secrets," she said, "but we loved each other before the mission."

There was scarcely time to check into the Ashoka Hotel before Andrian, the two Valyas, and Valery were off to the Raj Ghat to pay homage at the shrine erected where Mahatma Gandhi had been cremated in 1948.

The next day found Valya and her fellow cosmonauts at tea with the president and at luncheon with Prime Minister Nehru. On the following day, Mrs. Indira Gandhi presided at a luncheon for the Soviet celebrities. There Valya received the first of several Indian *saris* that she would be given during the visit—a beautiful gown of red and gold brocade. Valentina Bykovsky was presented a similar one. Much to Valya's annoyance, a television interviewer wanted to know about her romantic life rather than her voyage in space. The reporter asked her if she planned a honeymoon on the moon.

"I am spending it in India," she replied shortly.

Two days later, in Madras—where Valya greeted the crowds in a dark green *sari* with a *tilak* in the middle of her forehead—she told a news conference that Russian cosmonauts would land on the moon sooner than anyone thought. "The Soviet people's custom is to do a thing rather than talk about it," she said.

Andrian, in turn, told some Indian scientists that the USSR was planning a spaceship that would sustain life for three years—a period, they noted, that was equal to a round trip to Venus.

Two days later the group was in Hyderabad. On November 19 they landed in Katmandu for a two-day visit to Nepal. At the airport ceremonies, Valery and Andrian were presented with the quaint caps worn by the Nepalese.

They posed self-consciously for the photographers, wearing these incongruous hats with their formal air force uniforms. Valya, not knowing what to say to such an exotic audience, said rather lamely that some day she hoped to climb Mt. Everest.

On the following day, at the Royal Palace, they were received and decorated by King Mahendra. The visitors spent the night in the capital and left the next day to return briefly to India.

The pace became more and more tiring for Valya. After landing in Calcutta, there were the inevitable ceremonies at the airport and then a cruise on the Hooghly River. Following a breather at their hotel, the visitors attended a reception at Rabindra Stadium, which was sponsored by several women's committees. There Valya was greeted by the traditional rain of flower petals and blowing of conch shells. Valery got in the mood and began tossing handfuls of petals over his Valya, but Andrian dourly refused to engage in such frivolity. Another reception followed atop the New Secretariat Building, Calcutta's tallest. The hosts served tea and coffee, and Valya had a cup of black coffee. However, when they offered her *sandesh,* an extremely sweet native candy, she refused with what looked to some like a faintly green smile.

The day ended with still another reception, by the Indo-Soviet Cultural Society. Despite her obvious fatigue, Valya was the center of attention in her pearl-white dress and wreath of red and white roses. Among the gifts presented to her were a lotus flower and a casket of Darjeeling tea. The next afternoon Valya and her group left for a four-day visit to Ceylon.

They landed in Colombo on November 23. There was the usual tumult that they had come to expect. A quick trip through town, past waving, happy crowds, brought them to the hotel for a brief rest before beginning the arduous round of personal appearances. First, there was a luncheon with the minister for defense and external affairs, then a visit to the Soviet embassy, a children's rally, and another reception in Independence Hall. Finally, there was a reception given by the Soviet ambassador.

The next day found the two couples visiting a teachers' college, where they dutifully applauded students performing traditional native dances. On the following day the party began a lightning tour of several provincial towns.

The first stop was at Nuwara Eliya, some thirty miles south of the old capital of Kandy, almost in the center of the island. There they visited the famous Hakgalle Gardens and had a typical Ceylonese luncheon. Then the group traveled on to Kandy. Just outside the town they visited the Royal Botanical Gardens and the University of Ceylon. Like all visitors to Ceylon, the cosmonauts toured the temple in which a tooth of the Buddha is enshrined. After a reception at the Pushpanda Hotel, the four spent the night at the king's pavilion.

On November 26 the quartet headed north, stopping in Matale for a visit at Vijaya College. Most of the day was spent in the ancient capital of Anuradhapura. The highlight of their tour there included walking around the 2,000-year-old bo tree which was said to have grown from a branch of the tree under which the Buddha sat when he received his Enlightenment. From there the four went to see the Ruanveli shrine to the Buddha. Valya was amazed when she was told that it compared in size with the largest of the Egyptian pyramids.

Returning to Colombo, the cosmonauts made a state call upon the prime minister, Mrs. Sirimavo Bandaranaike. Then Valya and Valery submitted to the inevitable press conference at their hotel. Valya announced that she and the other cosmonauts who had been on a mission would not again go into space. She was to be proven wrong in the case of her new husband. She also said, significantly, "Prime Minister Nikita Khrushchev has no immediate plans for launching space flights to the moon." This statement conflicted with what she had said in Havana a month earlier, but in the meantime the party line had changed.

On November 27 the cosmonauts' plane touched down in Djakarta. A visit to Indonesia at the invitation of President Sukarno began with a walk down the familiar red carpet from the plane to the reception party. The cosmonauts were greeted by Deputy Premier Johannes Meimena and three small Russian children in cloth space suits who gave them each a bouquet of flowers.

That evening at a reception, President Sukarno awarded Valya and Valery the Star of the Republic Second Class. Andrian had been awarded the same medal after his *Vostok 3* mission. In return, Valya and Valery presented Sukarno with a model rocket bearing the pictures of the Soviet cosmonauts.

On the following day Valya and her group visited the Kalibata Heroes' Cemetery to lay wreaths on the graves of Indonesian dead. That night they appeared before a large rally to hear themselves and their country praised in socialistic terms. On November 29 the party flew to southern Sumatra. Landing at Palembang, where a crowd of thirty thousand greeted them, Valya and Valery were presented honorary wings by the air vice-marshal of the Indonesian air force. The activities in Indonesia ended with a quick trip to the isle of Bali, and the four then left the country for a brief visit to Burma.

In Rangoon the party was greeted by Colonel Hla Han, minister of health and education. Among the diplomatic delegations from the Communist and neutral countries awaiting them, the Chinese were conspicuous by their absence. After two exhausting days among the charming

Burmese, Valya literally had to drag herself up the stairs and into the plane for the trip to Moscow on December 11.

Valya thought that she would have a chance to rest and recuperate from the physical grind of the past month during the Christmas and New Year holidays at home. But there was the usual round of parties at Star Town.

By Christmas the rumors were all over Star Town and Moscow that Valya was pregnant. As if to disprove these rumors, Valya began the new year with a four-day trip to Ghana at the invitation of President Kwame Nkrumah. She was met by him and the Soviet ambassador at the airport at Accra on January 20, 1964. After inspecting an honor guard from the Ghanaian army, she appeared at a press conference at the Accra Press Club. On the following day she visited the Selwyn Market in Accra and distributed autographed pictures of herself and souvenir medals of her *Vostok 6* mission. The next day she was welcomed to the Achimota School by a contingent of students beating native drums in a traditional *akwaaba* ("welcome") . From there it was on to the site where the gigantic Volta Dam was being built with Soviet aid. That evening, after a very tiring day, she returned to Accra for one last ceremony. President Nkrumah presented her with the Order of Volta, the nation's highest award. On the following day she returned to Moscow.

By the time she had returned, everyone in Moscow knew that Valya was pregnant. On January 29 the government officially announced it and said that she would have to cancel a state visit to Italy. In New York, *Mademoiselle* magazine proudly named Valya as one of its ten annual merit-award winners for outstanding achievements in 1963, stating: "All in all, we feel that each of this year's winners is more than ordinarily individualistic in her talent."

Despite the pregnancy, Valya felt obliged to make a

Valya displays the Gold Medal of the British Interplanetary Society, awarded during her visit to London. (DAILY MIRROR PHOTO BY CYRIL MAITLAND.)

trip to England. After all, Queen Elizabeth II had wired her a personal message of congratulations during the second day of her mission, and she herself was expecting a baby.

Valya landed in London on February 4, and was met by photographers and reporters. She told them—not quite truthfully, considering her condition—"If I am picked for the first flight to the moon, I am ready to go." On the following day she had an audience with the queen, and graciously refrained from commenting on the elevator service in England, hers having gotten stuck between floors on the way to a television interview. While in London, Valya was awarded the Gold Medal of the British Interplanetary Society. On February 10 she returned to Moscow.

With the trip to England finished, Valya had to remain at home. A tour of Sweden and Norway in March was made by Valery and Yuri.

Ten days after Valya returned home, on February 20, Andrian received a new assignment. He took over command of the cosmonaut detachment from Gherman Titov, who in turn had replaced Yuri Gagarin, since Yuri had gone back on flight training for a new mission.

The next few months found Valya taking things easy at Star Town. She was studying at the Zhukovsky Air Force Engineering Academy, as were Yuri, Gherman, Valery, Pavel, and Andrian—and she kept up with her classes until a week before the baby was born.

As the time drew near for the baby's birth, Valya began to have some fears. Premature births were common in her family, and her doctor had decided that the baby would have to be born by Caesarian section.

The doctors at Star Town knew that Valya had been exposed to radiation during her *Vostok 6* mission. She had received a whole-body dose of about 44 millirads, or about the same amount of radiation she would receive during a chest X-ray at the doctor's office. During his *Vostok 3* mission Andrian had received a whole-body dose of 62 millirads. Since it takes a dose of some 100 rads directly into the gonads of the male to produce an increase in the frequency of abnormal sperm cells, Andrian was in little danger of transmitting genetic damage to his future children.

The baby was premature, but the Caesarian section presented no problems, as things turned out. Remembering what her mother had told her about her own name, Valya made sure her baby was named with no doubt. Yelena Andrianovna Nikolayeva was born at 2 A.M. on June 8. She weighed 6 pounds and 13 ounces and was 20½ inches long. Yelena was a perfectly healthy and sound baby girl. Andrian had canceled a tour of Yugoslavia in order to be in Moscow for the birth of his child. When the doctor called

The first photograph of Yelena Andrianovna Nikolayeva.

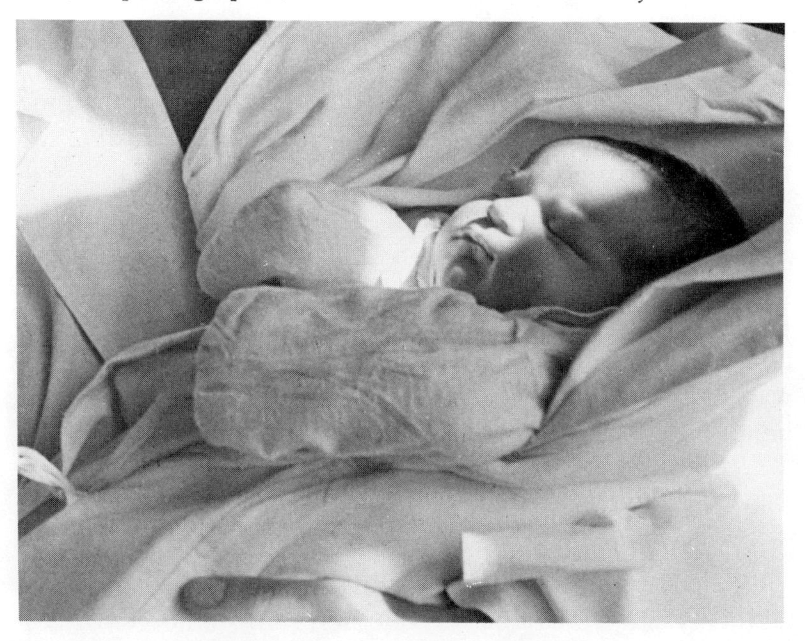

him and announced that he was now the father of a daughter, Andrian immediately drove to Moscow's Institute of Gynecology and Maternity and hurried to Valya's room on the third floor. The doctors made him put on a sterile gown, mask, and cap before he was allowed to visit Valya and to hold Yelena.

The news was not announced on the radio in Moscow until June 10; and when it was, throngs gathered in front of the hospital, shouting congratulations and waving gifts that they wanted to give to young Yelena.

Just to be on the safe side, and to give Valya a chance to recover from the surgery, the doctors kept her and Yelena in the hospital until June 23 before letting Andrian take them back to Star Town. During this period she was seldom without visitors. The room was literally filled with flowers and telegrams from the world over. Valya Gagarina visited her, bringing along her own Yelena and Galya. Both new grandmothers fretted about the number of people trooping in and out.

Once back in Star Town, Yelena occupied Valya's time until October, when her maternity leave was up and she returned to cosmonaut training status and her studies at the Zhukovsky Academy.

As with all young parents, time seemed to fly for Valya and Andrian. From the very beginning, it was apparent that Yelena was going to look more like her father than her mother. She had Andrian's dark eyes and hair, but from the way she grew, it was clear she would inherit her mother's height!

Life in the Nikolayev's apartment seemed to be in constant turmoil during Yelena's first year.

March 8, 1965, was a typical evening. Andrian and Valya had returned after a day of study at the Zhukovsky Academy and were debating whether to begin homework or

Valya and Yelena.

not. The two grandmothers had come in to baby-sit with little Yelena, although neither Andrian nor Valya had any plans to go out. Andrian decided to forgo the homework for the evening and see how the roll of movie film he had shot last week looked on the screen. Movie making had become a serious hobby with him in the last year. He began puttering with the projector on the dining room table.

The doorbell rang. Valya opened it to a stranger.

"Valentina Vladimirovna, I am Comrade Knipper, a special correspondent for the journal *Vodnii Transport,*" said the young man.

Valya told him to come in.

"I bring you greetings from the crew of the Baltic Sea freighter *Valentina Tereshkova,*" Comrade Knipper went on.

Valya and Andrian with Valya's brother Vladimir and her mother, Yelena Tereshkova.

"The crew of *what*?" Valya said dubiously.

Then she remembered the freighter she had visited in Havana a year before.

"Oh yes, by all means come in," she said, "but you will have to be quiet because Yelena is asleep."

After introducing him to Andrian, she sat talking to him while Andrian continued threading his latest epic into the projector.

'I see you have a piano," Comrade Knipper observed. "Which of you plays it?"

"I do," Valya replied, "but about as well as Andrian makes movies."

"I heard that," Andrian grumbled from the dining room. "Lights out," he announced. "It's show time."

It was a typical amateur's effort to record baby, mother, grandmothers, and father. People appeared without heads, baby wandered off the screen never to reappear, and father suddenly changed from a normal pace to an extremely slow one. Obviously the cameraman had inadvertently shifted from normal to slow motion.

Just as the film was over, the bedroom door opened and the two grandmothers came out.

"Yelena woke up, Valya," Yelena Fedorovna said.

While Comrade Knipper was describing activities of the crew aboard the *Valentina Tereshkova,* Yelena began crying. Valya was trying to pacify her when there was a knock on the door. Opening it, Valya saw a neighboring cosmonaut standing in the hall with an empty vase in his hand.

"Uh, Valya, I know you are busy," he whispered, "but I have a *friend* coming over, and I need some flowers to brighten up the place. Could you spare a few?"

Seeing that no one seemed really enthralled with stories about Soviet freighters on the run from Odessa to Cuba, Canada, and the United Arab Republic, Comrade Knipper decided it was time to leave.

"Well, Comrade Tereshkova, I can see you are busy," he said lamely. "I guess I had better be running along."

"Be sure that you say hello to the boys and girls on my ship," Valya said happily as she maneuvered him out of the door simultaneously with the young cosmonaut clutching his vase of flowers.

In April Andrian was promoted to lieutenant colonel at the same time as his good friend Pavel Popovich. In May Valya made her first trip to Paris. There she had received the Galabert Prize of the International Aeronautical Federation. Incredibly, it seemed to Valya, Yelena was one year old in the following month. In October Valya and Andrian had to leave Yelena behind while they made a two-week tour of Japan as guests of that nation.

The year 1966 opened with a tragedy for Valya and the entire cosmonaut detachment.

In mid-December 1965, Sergei Pavlovich Korolev had checked into a hospital in Moscow for a series of tests. He had a history of heart trouble and intestinal ailments dating back to 1957. However, he left the hospital on December 18, promising to return after the New Year's holiday for surgery recommended by the doctors.

As he was being prepared for the operating room, Sergei Pavlovich asked his surgeon, "How long do you think I can make it with this old heart?"

"Well," the doctor replied after hesitating, "I think you might be good for another twenty years."

"I would be satisfied with ten, although there is still a great deal to do," Korolev answered slowly.

The surgery on January 14 was long and involved. Sergei Pavlovich Korolev died during the operation.

Characteristically, he died thinking of the future. Sergei Pavlovich was a man of the future, who devoted his talents and energies to hastening a time in which human beings would move into space as life had eons ago moved from the seas onto land. When he died, the world lost the first of the doers who made realities of the dreams of visionaries such as Konstantin Tsiolkovsky, Robert H. Goddard, and Hermann Oberth.

Sergei Pavlovich's body lay in state in the Hall of Columns of the Trade Union House in Moscow, an out-of-the-way place for a state funeral. The cosmonauts, their families, and his associates from many years of rocketry filed by the coffin.

On January 18 a small crowd gathered in bitterly cold, wind-swept Red Square in front of Lenin's Tomb. Many of them had no idea why they were there or who S. P. Korolev was. They had been hurriedly assembled from nearby offices and told to attend.

They were mystified to see, on top of Lenin's Tomb, huddled against the wind knifing across Red Square, the top leaders of the Soviet Union, including several of the nation's most famous cosmonauts. These men extolled the many services of Sergei Pavlovich to his country and its space program.

Then Yuri Gagarin, whom they all recognized, stepped up to the microphone. "It was Sergei Pavlovich who saw us off into space and welcomed us back to our Soviet homeland," he said with tears in his eyes.

Then Sergei Pavlovich Korolev's ashes in a plain gray metal urn were placed in a small niche in the Kremlin Wall by a strange group of mourners. Some of them were cynical politicians who had refused to allow his name to overshadow their own; some were the men and women who had trusted their lives in space to his vision and skill.

Several days after Korolev's funeral, his successor was named. The new Chief Designer of Spacecraft was Mikhail K. Yangel. His academic and professional credentials were

A formal portrait of the Soviet cosmonauts. Front row, left to right: Andrian Nikolayev, Pavel Belyayev, Valentina Nikolayeva-Tereshkova, Valery Bykovsky, Pavel Popovich. Back row, Boris Volynov, Yevgeny Khrunov, Georgi Beregovoy, Vladimir Shatalov, Aleksei Yeliseyev, Aleksei Leonov.

certainly worthy—indeed they surpassed those of Sergei Pavlovich. However, Yangel was not given the authority or responsibility in the space program that Korolev had possessed. These were absorbed within the State Commission. Yangel assumed the title but in a narrow technical sense. He was concerned only with the design of manned spacecraft, not their overall manufacture, testing, and launching.

The space budget was constricted, too. The Americans were clearly going to win the race to put the first men on the moon. Despite these conditions, and with the same dedication that Sergei Pavlovich had shown, Yangel took up the challenge. At the time he could not know that he had only five years to tread the tightrope between party requirements and a budget that would have horrified Sergei Pavlovich.

In March Valya was promoted to the rank of major, and the increase in pay provided for a few luxuries such as a new television set. In the following month Yelena began talking, and Valya told the USSR about it very proudly in a television interview. Only two months later Yelena had her second birthday.

The following year was again one of tragedy.

For Valya it had a special significance. During the past year Andrian had been on a new assignment. He and Yuri were detailed as cosmonaut engineers on the team developing the new Soyuz spacecraft. It was for this project that Andrian had been sent to the Zhukovsky Academy. His schedule became increasingly more hectic as he got involved in the Soyuz program. He had to keep up in his studies at the Academy, work on the Soyuz design, keep up his jet-flight and parachuting proficiency, find time for physical training, and somehow manage to devote a few hours to Valya and Yelena.

Valya knew that eventually Andrian would be a crew member on the new spacecraft. By spring of 1967 it was obvious that the first manned flight in Soyuz was on the way. Three unmanned Soyuz ships, designated *Cosmos 140, 146, and 154,* had been launched into earth orbit on February 7, March 10, and April 8. Since everything had gone according to plan, the State Commission for Space Exploration now ordered the launch of the first manned Soyuz.

The flight was made on April 23, 1967. The cosmonaut chosen for the task was Vladimir M. Komarov, who had been the commander of the *Voskhod 1* mission in 1964. He had been in the first group of cosmonauts selected and had trained with Yuri and Andrian, serving as backup to Pavel Popovich on *Vostok 4.* That Komarov was given such assignments demonstrated the faith that Sergei Pavlovich had had in him and his abilities. Komarov had proven himself in 1962 when he underwent heart surgery to correct a defect that could have caused him to be dismissed as a cosmonaut; only five months after the operation, he was back up with his class in training.

The mission proved uneventful until the sixteenth and seventeenth orbits, when the Soyuz went out of control and began tumbling. Komarov managed to get it in the proper orientation for reentry on the eighteenth orbit and fired the retrorocket to brake the craft out of orbit. The tragedy occurred when the spacecraft had reached an altitude of a little over four miles. The hatch on the parachute compartment separated on schedule, but the risers on the parachute became tangled and the chute failed to open properly. Komarov was killed when the craft crashed into the ground.

Komarov was given a state funeral, and the urn containing his ashes was placed in the Red Banner Hall of the Central Club of the Soviet army so the public could pay its last respects. Tens of thousands streamed through the hall. Valya and Andrian took their turns in standing with the honor guard next to the flower-decked urn. On April 26 the urn was placed on a gun caisson drawn by an armored car, and the funeral procession moved out for Red Square, about a mile away.

Valya joined Andrian and the rest of the cosmonaut detachment among the mourners walking slowly behind the caisson. Immediately in front of them were Komarov's widow, Valentina, and their two children, fifteen-year-old Yevgeny and nine-year-old Irina. Valentina Komarova had not known anything about the *Soyuz 1* mission until after the launch. Behind the cosmonauts came the government and party officials, all slowly walking bareheaded through the dark, overcast spring day.

After the official obsequies were made, the cosmonauts assisted in placing the urn in a niche cut into the wall of the Kremlin, the most honored resting place in the USSR. As the plaque was placed over the niche, an artillery cannon fired three rounds in honor of Komarov's military rank.

On March 12 Valya's fellow citizens in Yaroslavl had elected her a deputy to the Supreme Soviet. She was one of

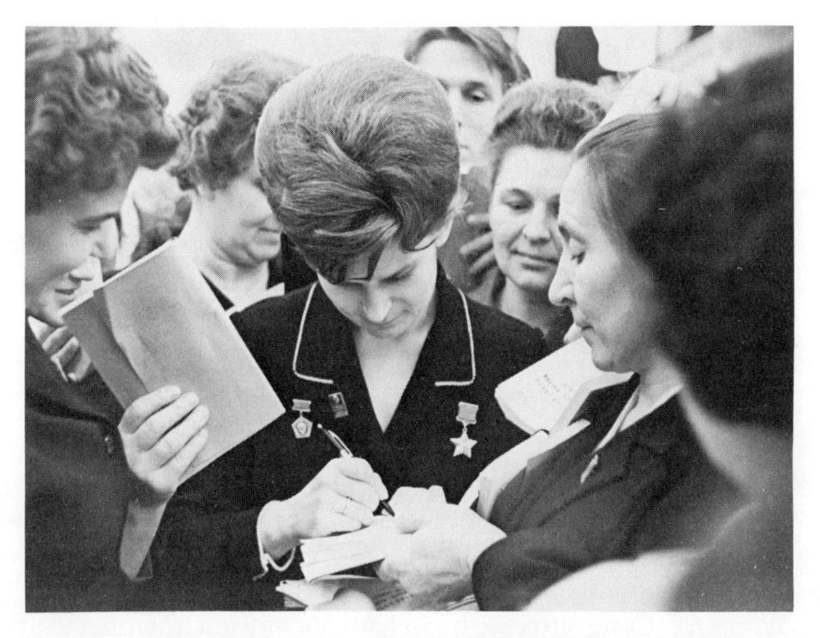

Signing autographs became a regular part of Valya's daily life.

1,962 women so elected. As such she participated in its meetings in October in the Palace of Congresses in the Kremlin, where she was mobbed by her fellow deputies, all wanting autographs.

Sea Gull Back to Earth

Honors, some with hard work attached, continued to accrue to Valya for several years after her *Vostok 6* mission.

In 1968 she was elected president of the Soviet Women's Committee, which had 40 million members! It was clear from the beginning that she intended to take her job seriously.

"The Soviet women have done me a great honor. I am proud of their trust and shall do everything in my power to strengthen the solidarity of the Soviet women and the women of the entire planet and their determination to work for peace, for the happiness of our children. But I shall remain in the contingent of cosmonauts. My profession is that of a space woman," she said in accepting the position.

Valya's office was on Pushkin Street in Moscow, and she spent several hours a week in it. A typical day found her in conference with a delegation of fellow Russian women planning meetings or discussing committee business. Invariably, the phone would ring.

"Hello . . . yes, it is I," Valya would say. . . . "What? You can't find anything to do? . . . Well, I will be home soon. I can't talk to you now; I am very busy. . . . Yes, yes, of course, I love you. Good-bye."

Her visitors nodded knowingly to each other. Little Yelena had called her mother.

In February of the same year, Yuri, Andrian, Gherman, Pavel, and Aleksei Leonov graduated from the Zhukovsky Academy, and things were looking up for the group of old friends. They were all on special assignments, preparing for greater exploits in space.

Then, on March 27, tragedy struck in Star Town.

Yuri Gagarin and Colonel Vladimir S. Seryogin, a test pilot and commander of the Soviet air force unit near Star Town, where the cosmonauts do their practice flying, took off on what was to be a routine training flight. Their specially modified two-seater Mig-15G roared off the runway and shot into the air. A few minutes later it developed engine trouble and crashed into a swamp, killing both men.

For two days the gray metal urns containing the ashes of Yuri and Colonel Seryogin lay in state in the Red Banner Hall of the Central Club of the Soviet army in Moscow. Each was enshrined in a floral bower, in front of which was a portrait. On cushions beneath the portraits were displayed the various medals each man had been awarded. Covering their common bier were flowers from the most important people in the government and the Communist party. Among them also were floral tributes from Yuri's comrades at Star Town and his old friends in the air force.

Tens of thousands of Muscovites filed past Yuri's urn. They were, for the most part, the common people who identified with the ex-foundryman with the friendly grin, who had been the first man to leave earth on man's journey ultimately into the farthest reaches of the universe.

A state funeral was held on March 30 for Yuri and Colonel Seryogin. The same armored car and gun carriage that had been used only a year before for Vladimir Komarov carried the two urns to Red Square for burial in the Kremlin Wall, an honor afforded very few citizens of the USSR. (Curiously enough, the Kremlin Wall is also the

The funeral of Yuri Gagarin and Vladimir Seryogin, March 30, 1968.

grave for one American. He was John Reed, a poet, writer, and revolutionist who died in Moscow in 1920.)

The hearse drove slowly through streets lined with citizens weeping openly. An honor guard of soldiers from the Moscow garrison marched slowly alongside the gun carriage. Directly behind it was Valentina Gagarina, accompanied and supported by Valya and Aleksei Leonov. Behind them came the other cosmonauts, all in military uniform or black suits with black armbands on the left sleeve. They were followed by the members of the State Commission for Space Exploration. Behind them came high government officials and party functionaries, all bareheaded despite the chill wind sweeping across Red Square.

The army band played Chopin's Funeral March, and its slow, sad melody echoed dismally off the gray, drab walls of the GUM department store and the rough, dark red bricks of the Kremlin Wall. During the obsequies Secretary Brezhnev, from atop Lenin's Tomb, at one point had to stop briefly when he saw that Valentina Gagarina had fainted. She was quickly revived by an air force doctor with a small cylinder of oxygen.

After the ceremonies were over, an event took place that had never occurred in the history of the USSR. For one minute, there was silence in the nation; every auto horn, factory whistle, work bell—indeed all activity—came to a halt as Russians honored the memory of one of their countrymen who had gained a unique place in the history of mankind.

Valentina Gagarina then leaned over briefly and kissed the picture of Yuri in full-dress military uniform that stood beside his urn. Brezhnev, Premier Aleksei N. Kosygin, and President Nikolai Podgorny, assisted by several cosmonauts, placed the two urns into their niches in the Kremlin Wall and fitted the memorial plaques over them.

In giving the cosmonauts' farewell to Yuri, Andrian said to the crowd below, brokenly and with tears in his eyes, "Our sorrow is deep and painful and incurable. But nothing will stop us on our way to conquer space. We are ready."

Thus was buried the first man to leave earth and to orbit it in a spacecraft.

The following year, 1969, opened dramatically for Valya. She was almost shot to death in the Kremlin by a man seeking to assassinate Secretary Brezhnev and other government officials.

The morning of January 22 had begun in a festive mood. Valya and Andrian, and most of the other cosmonauts, had gathered at Vnukovo Airport outside Moscow to welcome home the crews of *Soyuz 4* and *5*, the twin missions of January 14–17. Four cosmonauts had been involved in the mission, which saw two Soyuz spacecrafts dock in orbit and the cosmonauts of one transfer to the other by going outside in their space suits. The crewmen were Colonel Vladimir A. Shatalov, Colonel Boris V. Volynov, Colonel Yevgeny V. Khrunov, and Aleksei S. Yeliseyev, a civilian engineer.

After the traditional walk down the "one hundred steps to glory" on the red carpet and the report to Secretary Brezhnev, the four cosmonauts entered the first car in a line of twenty limousines that were to form the parade for the twenty-mile trip back to Moscow and Red Square.

Valya, Andrian, Colonel Georgi T. Beregovoy (who had piloted *Soyuz 3)*, and Aleksei Leonov entered a closed car following the open car with *Soyuz 4* and *5* crews, who were standing and waving to the crowds.

As usual on such occasions, the parade left the airport and drove along Lenin Prospekt and then turned on to

Dimitrov Street and moved slowly toward the bridge over the Moscow River. Along both sides of the street the usual crowds had gathered. The automobiles entered Borovitsky Square and turned right, through the Borovitsky Gate into the Kremlin.

Suddenly a man wearing a *militzia* uniform jumped out of the cheering crowd and began firing at Valya's car with a pistol in each hand.

Bullets crashed through the doors, shattering windows and spraying slivers of glass into the car. The driver slumped over the wheel. The car came to a stop. In the rear seat, Colonel Beregovoy's right hand suddenly grasped his face—and he withdrew it covered with blood. He had been hit by a piece of glass, but he was not wounded seriously. The driver, however, was mortally wounded and died later. Andrian immediately pushed Valya to the floor and crouched over her. A soldier on a motorcycle next to the car fell from it bleeding from a wound in the head. His cycle lay on its side, the motor still running and the rear wheel spinning.

Despite the incident, the ceremonies continued with only a ten-minute delay. Secretary Brezhnev and the four Soyuz cosmonauts mounted to the top of Lenin's Tomb, and as the usual speeches were made, only a handful of Muscovites below were aware of the shooting that had just occurred.

Two days later, at the news conference for the *Soyuz 4* and *5* crews, Aleksei Leonov humorously compared weightlessness to being shot at. "The weightless state is hard to explain and even more difficult to reproduce on earth," he said, "but this feeling of being shot at can easily be produced on earth and, incidentally, has been repeatedly described both in the Soviet Union and abroad."

It was later revealed that the assassin was a deranged

Soviet army officer from Leningrad and that he had mistaken Valya's car for the one in which Secretary Brezhnev and President Podgorny were riding.

May 31 was a beautiful spring day in Star Town. Valya put on her lieutenant colonel's uniform, assisted by Andrian, who showed her where all the medals and decorations went. She was getting dressed for graduation from the Zhukovsky Air Force Engineering Academy. It had been a long and difficult course for the girl who had already sacrificed so much of her time for her education. Now, however, she was a cosmonaut and an aerospace engineer as well. Andrian and Yelena sat quietly and very proudly among the guests as Valya accepted her diploma. After the ceremony, Andrian gravely kissed Valya and congratulated her formally.

Valya's graduation from the Zhukovsky Air Force Engineering Academy.

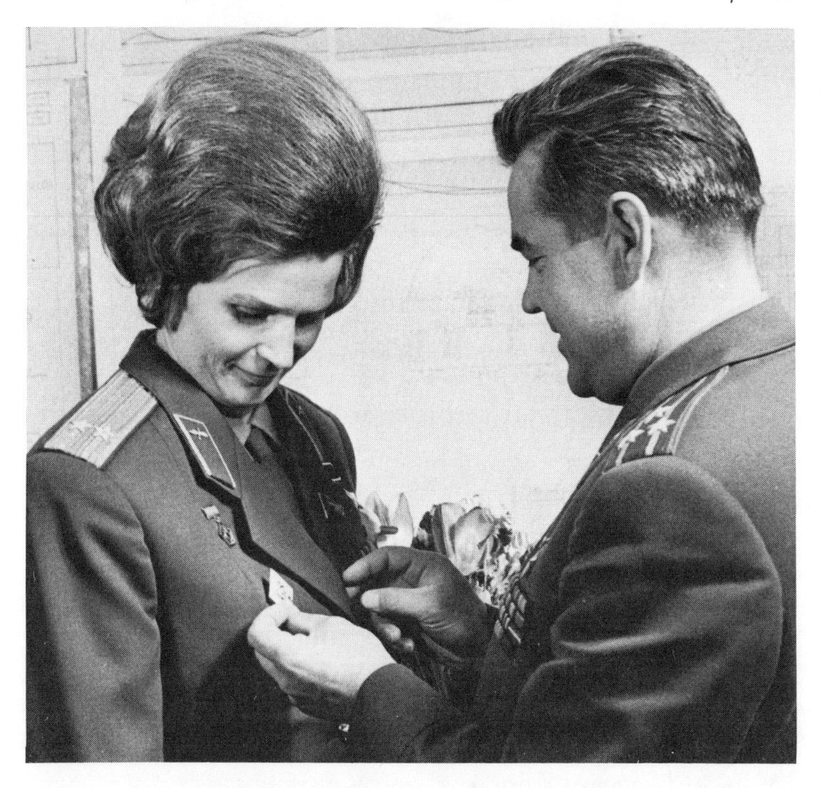

Andrian congratulates the new aerospace engineer.

Since Valya was still very much involved in what might be called a forerunner of the Women's Liberation movement, she received many requests to attend functions all over the world. As president of the Soviet Women's Committee, she flew to Finland on June 16 to address the Women's International Democratic Federation in Helsinki. The theme of the meeting was "The Role of Women in the Modern World." Valya was to present the Soviet Union's report, "Women and Labor."

By now an accomplished speaker, she approached the podium with confidence and self-assurance. "One of the most characteristic features of our time is the growth of female participation in industry in all the countries of the world," she said. "Women produce one third of the material wealth necessary for humanity. They are making a great contribution to the development of the world's culture and science. Now economic incentives are not the only reasons that women participate in labor. Women are coming to the conclusion that working in industry is the only way to attain equality, both in society and in the family." The audience applauded enthusiastically.

Valya had scarcely returned to Star Town when she was called upon to add to her other duties. She became the official hostess for the cosmonaut-training center. On July 5 the American astronaut Frank Borman, who had piloted *Apollo 8* around the moon at Christmas in 1968, paid a visit with his family. They were met in Moscow at the House of Culture by Air Marshal Pavel Kutakov and by Colonel Beregovoy, Gherman, Pavel, Valery, Andrian, and Valya. With Valya leading the way, astronaut Borman and his wife visited the Kremlin Wall, where they laid wreaths at the tombs of Gagarin, Komarov, and Korolev.

On visiting the museum at Star Town, Borman was so impressed with it that he spontaneously removed the wrist-watch that he had worn on his *Gemini 7* and *Apollo 8* missions and asked that it be placed in the display case with those of his Russian colleagues.

On the days when she was not playing hostess, Valya had a busy time. "We get up at 6:30 in the morning—we do physical exercise. After breakfast we have a conference and practical-training exercise. We work until late afternoon five or six days a week. Evenings are spent at home with my husband, daughter, and friends. Sometimes we go into Moscow. It's a wonderful life. I couldn't be happier," she told a friend.

In addition to her cosmonaut duties, she also had some domestic ones. "I am a good plain cook. No frills. I make the kinds of things people like—*borscht* [beet soup], *blini* [pancakes], and *pelmeni* [dumplings]—and my strawberry jam is out of this world," she told a reporter with an uncharacteristic lack of modesty.

Early in 1969, Valya's flight status was uncertain. A reporter asked her if she would go into space again.

"I don't know. It hasn't been decided yet; my greatest joy would be to become the first woman to set foot on the moon," she said.

In June, however, an event at the launching site at Tyuratam made Valya's hope of becoming the first woman to land on the moon a dim one indeed.

Standing on the launching pad at Tyuratam was the first flight vehicle of the Soviet's new three-stage rocket booster. It was powerful enough to send 40,000 pounds to the moon. It dwarfed the booster which was used with the Soyuz and had formerly been used with the Voskhod and Valya's *Vostok 6*. The new rocket had a lift-off thrust of nearly 10 million pounds, almost 2.5 million pounds more than the American Saturn 5 moon rocket produced.

During operations to make it ready for launching, something went wrong. There was a catastrophic explosion. The rocket disappeared and much of the equipment on the pad was destroyed or damaged. An American spy satellite orbiting over Tyuratam brought back pictures of the damage.

The Soviet manned lunar landing program had literally gone up in smoke and flames.

The booster explosion at Tyuratam was only one event during the year that caused a change in future Soviet space plans.

Because of a worsening political situation with Red

China, many of the large rockets allocated initially for the space program were recalled by the military. Thus launch schedules were upset. Also, some of the space-launch facilities at Tyuratam were turned over to the Strategic Rocket Forces during the Chinese border crisis.

In November Dr. Keldysh, president of the Academy of Sciences of the USSR, visited Stockholm. He freely admitted for the first time in public, "At the moment, we are concentrating wholly on the creation of large satellite stations. We no longer have any scheduled plans for manned lunar flights."

It was official. With the country's problems in foreign relations, the realignment of its rocket resources, and the success of *Apollo 11,* the Soviet Union had bowed out of the race for the moon. To make matters worse, the first unmanned launch of the new Salyut space station went awry and it failed to orbit.

January 1970 began sorrowfully for Valya and the cosmonauts of Star Town.

Pavel Ivanovich Belyayev, their comrade and first commander of the cosmonaut detachment at Star Town, died on January 10 at the age of forty-five, following an operation for ulcers of the stomach. Belyayev had been selected in the first group of cosmonauts with Yuri Gagarin and Andrian, and because of his experience and education had been made the first commander of the detachment. He held the position until he broke a leg while parachuting in 1960. He then turned over the leadership to Yuri.

Belyayev was held with great respect by his fellow cosmonauts, who addressed him as Pavel Ivanovich, although his close friends called him Pasha. He had proved his abilities as a cosmonaut when he piloted the malfunctioning *Voskhod 2* safely to earth in March 1965, after the mission in which Aleksei Leonov became the first man to walk in space.

Valya again found herself at a funeral, on hand to comfort Belyayev's two daughters, Irina and Ludmila.

The funeral recalled the sorrow she had known on two previous occasions. However, it especially brought back the memory of the funeral of Komarov, who had died testing the first Soyuz spacecraft. In five months Andrian would be making his second space flight, in *Soyuz 9*. Despite her cosmonaut's confidence in the reliability of the Soyuz, she still had a wife's apprehension of an imminent space mission.

In the months following Belyayev's funeral, forty-year-old Andrian began an intensive training program with one of the newer cosmonauts, thirty-four-year-old Vitaly Sevastyanov, a young Ph.D. in engineering. Their mission would be to set a record for time spent in orbit so that engineers could check the Soyuz under long-term use and the doctors could evaluate the long-term effects of space travel on cosmonauts.

Valya and Yelena, accompanied by Vitaly's wife, Alevtina, and their seven-year-old daughter, Natasha, went with Andrian and Vitaly to the airfield at Chkalovskoye to see them off for Tyuratam. With them were several of the cosmonauts who would remain behind to work in the mission-control center.

Since she could not go to Tyuratam with Andrian, Valya had to be content to watch the lift-off on closed-circuit television at Star Town.

The mission had been designed so that the Soyuz would initially be over the USSR at night and over the other half of the world during the day, so that certain astronomical, biological, and earth-resources experiments could be performed. This plan meant that the cosmonauts would have to reverse their daily rhythms of being awake and asleep. In other words, their "day" would begin at about 1 P.M. and end at 5 A.M.

In order to accomplish this shift in routine, Andrian and Vitaly spent more than two weeks at Star City following a schedule in which they stayed up at night and slept during the day. Andrian even went fishing one "day" and caught a 48-pound sheatfish!

Valya and Andrian had agreed that she should stay in Star Town rather than go to Tyuratam for the launch and the mission. She had Yelena to look after, and she had her own training and instructing to do. In addition, she had another hostess duty coming up.

On May 31 the American astronaut Neil Armstrong, the first man to land on the moon, nearly a year earlier, visited Star Town.

He was introduced by Colonel General A. M. Yefremov, the deputy commanding general of the Soviet air force, in the auditorium. Needless to say, the place was packed with cosmonauts, wives, children, scientists, engineers, and important people from Moscow. Armstrong showed a film of his *Apollo 11* mission and accompanied it with a professional and often witty commentary which brought laughter from the assembled Russians on several occasions. Indeed, Armstrong's stocky build, fair hair and easy smile reminded Valya and her comrades very much of Yuri Gagarin.

After his talk, Valya presented him with a large bouquet of flowers, and General Beregovoy gave him a model of two Soyuz spacecraft docked. In accepting these gifts, Armstrong told the group that he had left medals on the moon in memory of Yuri Gagarin and Vladimir Komarov. Then he asked Valentina Gagarin and Tatyana Belyayeva to join him. He presented them with duplicates of the medals and gave General Beregovoy a set for the museum.

An American visitor arrived just as Andrian's Soyuz 9 *was scheduled for launch. Neil Armstrong's easy smile reminded Valya and the other cosmonauts of their own space pioneer, Yuri Gagarin.*

Following these ceremonies, Valya accompanied Armstrong on a tour of Star Town. She was especially touched as he wrote in the visitors' book in Yuri Gagarin's study: "He called us all into space." She also looked on as he flew a docking mission in the computer-controlled Soyuz spacecraft simulator.

The following evening, Armstrong was invited to

dinner by General Beregovoy, who told him that *Soyuz 9* was scheduled for launch that night.

"Neil, two of our boys are lifting off tonight," he said after dinner. "Would you like to watch it on television?"

Exactly at 10:09 P.M., on June 1, 1970, *Soyuz 9* lifted off from the pad in the USSR's first manned launching at night.

The mission went according to schedule, and Andrian and Vitaly quickly settled into their routine.

On June 4 Vitaly became an orbiting cameraman and trained his camera on a reticent Andrian. Unaccustomed to the role of commentator, Andrian proved that the reporters of the USSR had nothing to fear. Their jobs would not be in jeopardy after he returned to earth. Very stiff and ill at ease, he gravely explained to his Russian viewers what the functions of various controls in the spacecraft were. In closing, he promised to explain more about the control system at a later date.

With the television show over, cosmonaut Viktor Gorbatko, the spacecraft communicator in mission control, told Andrian that he had just received a telegram for him from the city of Yaroslavl. It was from a candidate who was running for a second term as deputy to the Supreme Soviet. The telegram was, of course, from Valya, who took time out from her politicking to wire, "We send you greetings, darling, from Yaroslavl."

As the mission progressed, it became clear that Andrian and Vitaly could get by on only six hours of sleep a day. Thus they spent eighteen hours a day working. The doctors were especially interested to learn this piece of news since they were curious to know how well the daily rhythms of the cosmonauts had adjusted. The biological telemeters indicated that both cosmonauts were doing fine living by an inverted clock.

June 8 was Yelena's birthday, and Valya let her stay up

several hours past her bedtime. She and Andrian had a special present for Yelena—a "happy birthday" greeting from Andrian as he orbited over Moscow at 10:20 P.M.

The television screen flashed on with a picture of Andrian waving into the camera. "Happy birthday, Yelena," said Andrian with a big smile. "I can't believe you are six years old today!"

Yelena was pleased with her surprise, but she had a question. "When will you be home, Papa?" she asked into a microphone.

"In just a few more days, dear," he replied.

A few days later Andrian and Vitaly decided to swap bedrooms. Andrian moved into the orbital compartment of the two-room Soyuz, while Vitaly took the command compartment. The reason was very simple and unscientific. Andrian liked to sleep in a room where the temperature was about 75° F., whereas Vitaly preferred a cooler bedroom.

On June 10 mission control decided that the two cosmonauts could have a day off. They deserved, and needed, a day of rest.

"Since we haven't any place to go," quipped Vitaly, "how about a game of chess with you guys on the ground?"

For a few hours, the two cosmonauts played games with General Kamanin and their fellow cosmonaut Viktor Gorbatko. Then Andrian took the opportunity to catch up on his reading, and both cosmonauts pitched in to do housecleaning. The first thing they did was vacuum-clean the air in their cabin! Using a small vacuum cleaner, they went carefully over the two compartments of the spacecraft, sucking up crumbs of food, drops of water, hair, dandruff, small bits of dust and dirt, and the other debris that inevitably collects with time in the spacecraft and floats in its atmosphere.

On the night of June 16, the day Valentina had gone

into space, Andrian had another special television broadcast for mission control. Valya and Yelena were there for it, too.

"Valya," he said with a smile, "happy seventh anniversary!"

"Anniversary?" Valya answered. "Have you forgotten when we were married?"

"Married?" he replied. "Have you already forgotten June 16, 1963?"

On June 19, a little after 2 P.M. and during its two hundred and eighty-eighth orbit, *Soyuz 9* fired its retrorocket engine. The spacecraft landed at 2:54 P.M. outside the village of Intumak, some forty-seven miles west of Karaganda, in the coal-mining region of Kazakhstan, the landing zone for most Soviet manned space flights. Andrian and Vitaly had spent 17 days, 16 hours, and 59 minutes in space. Four helicopters were circling the landing area when they saw the Soyuz descending by its parachute. The helicopters with medical personal and newspaper reporters landed almost at the same time.

The hatch popped open, and the doctors and medical technicians watched as Andrian and Vitaly struggled with their seat belts. Both then tried to get out of the contoured seats, but the effort was beyond them.

"I feel heavy as lead," said Andrian.

"My space helmet feels like a deep-sea diver's helmet," Vitaly added.

The two had to be lifted from the Soyuz by the recovery team.

Eighteen days in weightlessness had affected both of them in this curious way. Once out of the Soyuz, their clothes felt like weights hanging on their bodies. But one thing that weightlessness had not affected was their appetite. Vitaly wanted a good bowl of cold borscht, while Andrian craved some radishes!

Andrian emerges from Soyuz 9 *after 17 days in space.*

Unlike cosmonauts returning from earlier missions, Andrian and Vitaly had no triumphal return to Moscow and no red carpet with its "one hundred steps to glory." They went immediately to the hospital at Star Town. They simply were not in physical shape to take part in the usual ceremonies at Vnukovo Airport.

Once in the hospital, they underwent an extensive physical examination. The doctors found that 424 hours and 59 minutes in space had left the two men pale, extremely tired, and somewhat inattentive. Both had lost weight. In this condition, the doctors felt, they would be especially susceptible to common earth diseases such as colds and flu. So all communication between the cosmonauts and scientists, engineers, and reporters was conducted with Andrian and Vitaly in a room with a glass window between them and their interrogators.

Valya, Yelena, and Andrian at home in Star Town.

For ten days the two remained in the hospital, undergoing tests and readjusting to the normal pull of gravity. Then they were given a few days off to get reacquainted with their families and prepare for their first formal appearance in public.

On the evening of July 3, Valya put on her best evening gown and arranged for a baby sitter. She and Andrian had an important date in St. George's Hall of the Grand Palace of the Kremlin. They had both been there before, on two occasions when they were being honored. This time Andrian was being honored again for his *Soyuz 9* mission. He was promoted to major general and given his second Gold Star of the Hero of the Soviet Union, and Vitaly was given the title Hero of the Soviet Union and the Gold Star as well as the Order of Lenin.

In the following October, Andrian and Vitaly returned Neil Armstrong's visit. They toured the United States for ten days, visiting the Johnson Space Center in Houston, Texas, and the Marshall Space Flight Center in Huntsville, Alabama.

At the Johnson Space Center, on October 22, Andrian tried his hand at a simulated landing on the moon in an Apollo lunar module. He stood inside the lunar-module simulator with American astronaut David Scott, who would later land on the moon as commander of *Apollo 15*. Unfor-

A reunion of cosmonauts and astronauts: Astronaut Russell Schweickart, Andrian Nikolayev, Vitaly Sevastyanov (in a U.S. Apollo space suit), and astronaut Edwin Aldrin, at the Marshall Space Flight Center, Huntsville, Alabama, October 20, 1970.

tunately, Andrian did not make it to the moon in one piece! However, it was not a true test of his lunar-landing ability. His translator could not keep up with the English being called out by Scott as the craft in theory approached the moon's surface. Also, Andrian was not used to the type of instrument display in the simulator. He and Scott "crashed" onto the moon, but walked away from it smiling.

The last stop on the cosmonaut's tour of the United States found them at Disneyland in California. Both agreed that it was the place where they had the most fun during the trip.

The year 1970 ended on an encouraging note for Valya and her fellow cosmonauts. On December 2 the new Salyut space station was successfully orbited. It was given the designation *Cosmos 382* to keep its purpose a secret.

The following year was marked by events that discouraged the cosmonauts and technicians at Star Town. On April 19 the first announced Salyut space station successfully achieved orbit, and four days later *Soyuz 10,* with cosmonauts Vladimir Shatalov, Aleksei Yeliseyev, and Nikolai Rukavishnikov aboard, docked with it. However, something went wrong with the procedure, and the crew could not enter the Salyut. They returned to earth on April 24. On June 6, *Soyuz 11* was launched from Tyuratam. This time things went better. Cosmonauts Georgi Dobrovolsky, Vladislav Volkov, and Viktor Patsayev docked with the Salyut and entered it. They stayed aboard until June 29. While they were aboard, however, they received some bad news from Tyuratam. The new big booster had failed again, as it had two years earlier. In a way, this news was a prelude to an even greater tragedy.

On June 29 the three cosmonauts reentered the *Soyuz 11*. Dobrovolsky radioed that everything was working fine and he was proceeding as planned.

"You look down there and you get homesick. You want some sunshine, fresh air, and you want to wander in the woods," radioed Volkov.

The Soyuz reentered the atmosphere, and the radio transmissions ceased as the layer of ionized air formed around it.

This loss of communication was normal. When the blackout was over, mission control called the spacecraft.

"Amber, this is Dawn; come in," called the communicator over and over.

There was no reply.

The medical helicopter landed at almost the same time as the *Soyuz 11,* which touched down perfectly after its parachute deployed as planned at an altitude of five miles. The craft had landed in the usual zone near Karaganda.

However, the crew hatch was still closed.

Medical personnel and technicians quickly opened it, and sunshine, fresh air, and fragrance from the nearby forest filled the bell-shaped spacecraft. Seated peacefully in their contoured couches were the three cosmonauts. Dead.

Later their death was ascribed to the failure of the valve between the orbital compartment and the descent stage of the Soyuz. When these two units were separated during the undocking, the air in the spacecraft quickly rushed out through the leak into the vacuum of space. The three men died within twelve seconds or so, painlessly and quickly.

The state funeral in Red Square for cosmonauts Dobrovolsky, Volkov, and Patsayev was held on July 2. Under bright, sunny skies, thousands of sorrowing Muscovites gathered to watch the three gun carriages towed by armored cars move slowly past them, each with a flower-draped, gray metal urn in the center of it.

On this occasion, there were three new niches in the

Kremlin Wall beyond those of Gagarin, Korolev, and Komarov. As the urns were placed in them by the cosmonauts, assisted by Brezhnev, Kosygin, and Podgorny, Valya placed her arms around young Maria Dobrovolsky and comforted her, even though she could scarcely see through her own tears.

On a brighter note, in the following month both Valya and Andrian received an honor given to very few people on earth. A pair of craters on the moon were named for them by the nineteenth General Assembly of the International Astronomical Union, meeting in Brighton, England.

In 1972 another rash of technical problems arose to plague the engineers who were working on the problem of redesigning the balky valve in the Soyuz spacecraft. On June 26 an attempt to check out the new Soyuz in orbit failed, and the failure was disguised for the record books as *Cosmos 496*.

On the bright summer morning of June 30, a convoy of black limousines from Moscow sped through the green gate at Star Town. An important visitor from a foreign country had come for a tour, and Valya was on hand to greet someone she had not seen since 1963. Prime Minister Fidel Castro of Cuba, accompanied by Secretary Brezhnev, was visiting Star Town on his ten-day tour of the USSR. With Valya were General Shatalov and General Beregovoy as well as Andrian, Pavel, and Valery. Also in the party to greet the premier were some of the younger cosmonauts—Boris Volynov, Yevgeny Khrunov, and Georgi Shonin.

Dutifully Premier Castro laid a bunch of flowers at the foot of the statue of Yuri Gagarin. He then accepted General Shatalov's invitation to "fly" the Soyuz spacecraft simulator. Clad in his green combat fatigue uniform and smoking a cigar, the bearded premier quickly proved that his aptitude as a cosmonaut was not of a level to merit an invitation to stay in Star Town and take up a new career in space. Still, Castro was impressed. "These are unforgettable minutes for me," he said.

After the premier's visit, bad luck returned to the engineers of Star Town. At the end of July, a Salyut space station did not achieve orbit when its booster failed to operate properly. Several months later, the big booster finally lifted off successfully, but the first stage soon malfunctioned. The second stage never had a chance to ignite.

In her capacity as president of the Soviet Women's Committee, Valya was often called upon to visit other nations. In January 1973 she returned to India. She had been invited by both the government and the National Federation of Indian Women to celebrate Republic Day.

"We want peace, happiness for our children, and a clear sky over earth," she said on arriving.

She championed women's rights, and then went even farther. Valya came out strongly for children's rights. In answer to a query concerning what she wanted Yelena to be when she grew up, Valya paused only briefly before saying that the choice was her daughter's. "Perhaps she will want to be an engineer. Who knows?" she added.

From reporters in Bombay she faced the inevitable question of whether she had any plans to go into space again.

"It is my work. I am working for it. But no definite date has been set for it," she said carefully.

The newsmen misinterpreted her statement and rushed off to wire their newspapers that Valya was training for another space flight!

Asked if Indian women could ever hope to become cosmonauts, Valya had a ready answer. "Why not? Indian women can certainly aim for space flight. But it requires

In the midst of her many duties, Valya finds time for music.

some special and complicated training. She must be at least a pilot," she said.

In Bombay, on February 1, Valya met the press in the Russian consulate. She told them, quite truthfully on this visit, that her government had no plans to send men to the moon or women into space in the near future.

A Bombay reporter, obviously unaccustomed to interviewing Russians, was something less than kind in his coverage of her visit. "Valentina . . . parried questions of newsmen with long-winded answers rendered longer in translation from Russian," he fumed.

Today, as each new space mission takes place, Valya joins in a tradition that was set in Star Town before she arrived. When a man goes into space, the wives of the cosmonaut detachment all visit his wife on launch day or night, and when a safe return is announced, they pay her another visit.

But not everything has remained the same at Star Town. It has changed considerably since Valya arrived there in 1962 and since Andrian first entered it in March 1960.

It was still growing and expanding in 1974. Its three thousand inhabitants live in a complex of ten modern apartment houses. Near the familiar green gate the portraits of cosmonauts who have flown missions have been replaced by large black marble tablets. The *Soyuz 11* tablet has the names Dobrovolsky, Volkov, and Patsayev engraved in gold. A little way beyond, in the center of the apartment complex, is a huge stone statue of Yuri Gagarin. The area around Lenin Square now features a hotel, a community center where visitors are entertained, and a shopping center. In 1969 the elementary school had to be enlarged to accommodate the growing number of children in Star Town.

Birches like those Valya swung on as a girl grow near the cosmonauts' quarters in Star Town where she lives today.

The hospital and medical facility is larger, also. There is the special quarantine laboratory where Andrian and Vitaly spent their ten days following the *Soyuz 9* mission. An Olympic-size swimming pool is now a feature of the athletic facility, and there is also a new football field. A planetarium is now available for training cosmonauts, astrophysicists, astronomers, and the schoolchildren of Star Town. Even the central heating plant has been modernized to ensure no more freezing during the bitterly cold winters.

The little space museum off Lenin Square that Yuri showed Valya on her second day at Star Town is no longer a little one. The curator today is Tatyana Belyayeva, the widow of cosmonaut Pavel Belyayev, who died in 1970. Some of the original displays are still there: *Vostok 1*, Yuri Gagarin's space suit, the life-support pack used by Aleksei Leonov during the first space walk in 1965. In the various cases are the logbooks, map cases, pens and pencils, and wristwatches of the early cosmonauts.

However, there are now some three thousand gifts that have been presented to the various cosmonauts from people and governments all over the world. As more arrive after each mission, there is less and less space for them.

Now on view for the visitors are a Salyut space station and the *Soyuz 2* spacecraft. Visitors note with interest that the instructions for opening the hatch on the Soyuz are printed in Russian and English!

One of the most popular and revered exhibits in the museum today is the Yuri Gagarin Room. It is, in fact, his old office. The room has his desk just as it was the day he died. The papers on it are covered by a plastic case to discourage souvenir collectors. The wall behind it is covered by a huge map of the USSR, and the clock on the wall is set at the time of his death. His overcoat hangs on the clothes tree, and there is a picture of Smolensk, his hometown, on the wall.

Star Town has become such a popular place for official visitors to the USSR that it has become necessary to establish a guide unit to show them around. The sight of a group of visitors with a guide in the lead is a familiar sight in Star Town today. It is also a sight that sends well-known cosmonauts, including Valya and Andrian, scurrying down side streets to avoid the autographs and cameras!

The training facilities for the cosmonauts have been expanded, too. There is now the Volga docking simulator, which trains new cosmonauts in the docking of the Soyuz spacecraft to another Soyuz or to the Salyut space station. The Soyuz spacecraft simulator is connected to a computer-driven display that provides a view of earth or sky as the cosmonauts see them at any time during their mission. The full-scale replica, or mock-up, of the Salyut space station is kept up to date as changes are made in the flight model. Thus the cosmonauts can keep current without having to visit the manufacturing plant.

A new building and centrifuge are now available to provide training for both Soviet cosmonauts and American astronauts scheduled for the joint Apollo-Soyuz space mission in 1975. Nearby is a building for the simulator that trains the same cosmonauts and astronauts in the docking of the Soyuz and Apollo spacecraft for the mission.

On the shore of the man-made lake, a new high-rise apartment house is available for the American astronauts and space engineers who visit Star Town for training and coordination of the Apollo-Soyuz mission.

While there were some fifty cosmonauts in training in 1974, not all of them lived in Star Town. About a third of the corps were civilian scientists or engineers, who worked in laboratories or factories in or near Moscow. They lived close to their work and commuted to Star Town for training. However, there were no women in their number.

Concerning this situation, General Shatalov said, "In the future, especially in long space flights, women will take their places in our spaceships. They will be going as specialists. We have many women who specialize in meteorology and medicine, which are professions necessary in orbital stations."

When these women do take their place in space in the coming years, they undoubtedly will attend the Yuri A. Gagarin Training Center for Cosmonauts at Star Town. If the time is not too far in the future, they will be greeted and taught by Valentina Vladimirovna Nikolayeva-Tereshkova, who, as Sea Gull, blazed the trail for them into space in a quaint old spacecraft named *Vostok 6* on June 16, 1963.

Bibliography

von Braun, Wernher, and Ordway, Frederick I., III. *History of Rocketry & Space Travel.* New York: Thomas Y. Crowell, 1966.

Daniloff, Nicholas. *The Kremlin & the Cosmos.* New York: Alfred A. Knopf, 1972.

Gatland, Kenneth, *Manned Spacecraft.* New York. Macmillan, 1967.

J. L., "Study of a Space Child." *Saturday Review.* Vol. 50, No. 48, December 2, 1967, pp. 72–73.

Lebedev, L.; Lukyanov, B.; and Romanov, A. *Sons of the Blue Planet.* TT 72-52004. Washington: National Aeronautics and Space Administration/National Science Foundation, 1973.

Life. "She's the Sweetheart of the Soviet Sky." Vol. 55, No. 11, September 13, 1963, p. 43.

LUCE, CLARE BOOTHE. "An Amateur Chutist Takes a Giant Leap but Some People Never Get the Message." *Life.* Vol. 54, No. 26, June 28, 1963, pp. 31–32.

The McGraw-Hill Encyclopedia of Space. New York: McGraw-Hill, 1968.

Mademoiselle. "The Individualists: Mlle.'s Annual Merit Awards." January 1964, pp. 72–73.

Newsweek. "His and Hers." Vol. 62, No. 1, July 1, 1963, pp. 46–47.

PETROV, G. I. *Conquest of Outer Space in the USSR.* TT 72-52004. Washington: National Aeronautics and Space Administration/National Science Foundation, 1973.

PETROVICH, G. V., ed. *The Soviet Encyclopedia of Space Flight.* Moscow: Izdatelsvo Mir, 1969.

RIABCHIKOV, EVGENY. *Russians in Space.* New York: Doubleday, 1971.

SHELDON, CHARLES S. *Review of the Soviet Space Program with Comparative United States Data.* New York: McGraw-Hill, 1968.

SHELTON, WILLIAM. *Soviet Space Exploration: the First Decade.* New York: Washington Square Press, 1968.

STEVENS, EDMUND. "Comely Cosmonaut." *Ladies Home Journal.* Vol. 80, No. 7, September 1963, pp. 60–62.

STOIKO, MICHAEL. *Soviet Rocketry, Past, Present, and Future.* New York: Holt, Rinehart & Winston, 1970.

TERESHKOVA, VALENTINA. "Three Days in Outer Space." *The Saturday Evening Post.* Vol. 237, No. 1, January 4-11, 1964, pp. 61–63.

——— "Women in Space." *Impact of Science on Society.* Vol. 20, No. 1, January-March 1970, pp. 5–7.

Time. "Women Are Different." Vol. 81, No. 26, June 28, 1963, pp. 26–29.

References

ASTASHENKO, P. T. *Akademik S. P. Korolev.* Moskva: Izdatelsvo "Mashinostroenie," 1969.

BORISENKO, IVAN G. *Rekordnii Polet "Chaika."* Moskva: Izdatelsvo DOSAAF, 1966.

DOKUCHAYEV, YURI. *Kosmonavt i Evo Rodina.* Moskva: Izdatelsvo Novosti, 1967.

KISELEV, A. N., and REBROV, M. F. *Pokoriteli Kosmosa.* Moskva: Ordena Trudogo Krasnogo Znameni Voennoe Izdatelsvo, 1971.

Pokorennie Kosmosa. Moskva: Izdatelsvo "Mashinostroenie," 1969.

ROMANOV, A. *Konstruktor Kosmicheskix Korablei.* Moskva: Izdatelsvo Politicheskoi Literaturi, 1969.

——— *Kosmodrom, Kosmonavti, Kosmo.* Moskva: Izdatelsvo DOSAAF, 1966.

TERESHKOVA-NIKOLAYEVA, VALENTINA. "Iz Kosmosa na Parashute." *Aviatsita i Kosmonavtika.* No. 3, 1967.

——— *Vselennaya–Otkri tii Okean.* Moskva: Izdatelsvo "Pravda," 1964.

TITOV, GHERMAN. *Golubaya moya Planeta.* Moskva: Ordena Trudogo Krasnogo Znameni Voennoe Izdatelsvo, 1973.

Uspekhi SSSR v Issledovanii Kosmicheskovo Prostransva. Moskva: Izdatelsvo "Nauka," 1968.

VOSKRENSENSKII, A. D.; GAZENKO, O. G.; MAKSIMOV, D. G. *Vtoroi Gruppovoi Mosmicheskii Polet i Nekotoriye Itorgi Poletov Sovetskikh Kosmonavtov na Korabiyakh "Vostok."* Moskva: Izdatelsvo "Nauka," 1965.

Index

Page numbers in italics refer to illustrations.